T0190119

Textile Science and Clothing Technology

Series editor

Subramanian Senthilkannan Muthu, Kowloon, Hong Kong

More information about this series at http://www.springer.com/series/13111

Subramanian Senthilkannan Muthu
Editor

Sustainable Innovations in Recycled Textiles

Editor
Subramanian Senthilkannan Muthu
Kowloon
Hong Kong

ISSN 2197-9863 ISSN 2197-9871 (electronic)
Textile Science and Clothing Technology
ISBN 978-981-13-4175-5 ISBN 978-981-10-8515-4 (eBook)
https://doi.org/10.1007/978-981-10-8515-4

Softcover re-print of the Hardcover 1st edition 2018

Printed on acid-free paper

This Springer imprint is published by the registered company Springer Nature Singapore Pte Ltd.
part of Springer Nature
The registered company address is: 152 Beach Road, #21-01/04 Gateway East, Singapore 189721, Singapore

This book is dedicated to:
The lotus feet of my beloved
Lord Pazhaniandavar
My beloved late Father
My beloved Mother
My beloved Wife Karpagam and
Daughters—Anu and Karthika
My beloved Brother
Last but not least
To everyone working in the global apparel
sector to make it SUSTAINABLE

Contents

Recycled Fibres

P. Senthil Kumar and P. R. Yaashikaa

Abstract The present advanced world has created an expanded interest in innovation and production so quickly in all parts of the required living products. With a specific end goal to meet all the required requests, over creation and use of all assets appear insufficient. Accordingly, the expanding interest in producing tremendous textile garments isn't just in light of interest for more populace but on the other hand, it's changing new form propensities too. Enhancing crude material abuse has turned into an essential test confronting the logical and mechanical group. Textile production wastes are difficult however unavoidable side-effects in many assembling processes and are regularly underestimated. Nonetheless, in the event that one can change over such squanders into valuable item monetarily, there will be an incredible commitment to the market. As the material, attire, and retail businesses move to end up noticeably more maintainable, a zone of intrigue is the utilization of recycled fiber, yarn and texture in the improvement and generation of new products. The choice to utilize reused materials in items must happen at the outline and product improvement and proceed all through the assembling forms. Textile fibres can be recycled mechanically or chemically. Accumulation and Processing are two noteworthy advances engaged with recycling textile fibres. The cycled fibres find immense application in different fields, for example, manufacturing of garments using recycled fibres, home and fashion products, and electrical products and so on.

Keywords Recycled fibres · Open and closed-loop recycling · Characteristics
Methods of recycling fibres · Applications

1 Introduction

Textile industry is confronting few difficulties to create non-perilous strong material waste that can be recycled as required by customers and neighbourhood experts. It

P. Senthil Kumar (✉) · P. R. Yaashikaa
Department of Chemical Engineering, SSN College of Engineering, Chennai 603110, India
e-mail: senthilchem8582@gmail.com

S. S. Muthu (ed.), *Sustainable Innovations in Recycled Textiles*, Textile Science and Clothing Technology, https://doi.org/10.1007/978-981-10-8515-4_1

is unavoidable that a more economical practice in material assembling is required towards making zero waste by recycling and reusing textile waste materials into usable items, for example, building materials. Recycling rehearses, thus, would decrease the reliance on essential assets and guarantee a more supportable way to deal with living.

As requirement expanded, the assembling business advanced and difficult work frameworks were supplanted by automated assembling. This enabled textile material to be delivered less expensive, snappier and in huge qualities. This has brought about an excess of mass delivered shoddy and regularly low quality items and expansive volume of textile material filaments squander that has restricted end utilize applications. Textile material generation squanders cover every one of those crude materials which are either accumulating or being utilized as a part of the material business, for example, creation leftovers, squanders from fiber and fiber make, squanders from spinning, weaving, sewing and making-up and in addition reprocessed materials. Before, squander delivered during the production process (spinning, weaving and sewing) was regularly gathered and sold to the waste spinner at generally low costs. Certain mixing of squanders with great materials was essential keeping in mind the end goal to update the nature of waste yarns created and anticipate unreasonable end breakage rate at the time of spinning (Bhatia et al. 2014).

A procedure for utilizing recycled squander material for delivering a material item is given. This procedure can incorporate gathering diverse classes of waste material from an assortment of material arrangement forms. The procedure can likewise incorporate choosing particular classifications of waste material to be mixed together relying upon the last material product to be delivered. Recycling of material waste gives fiber a moment life in a revived life cycle and in this way expands the aggregate estimation of that recycled fiber. Still the greater part of recycled fibres wind up in low esteem items, so the advancement of new higher esteem items from recycled fibres will support usage of the waste filaments and add to the future manageability of industry. Today, recycling has turned into a need not just in light of the lack of anything yet additionally to control contamination. There are three approaches to decrease contamination. One is to utilize fresher innovations that pollute less. The other is to viably treat the textile effluent with the goal that the last emanating fits in with the normal standards. The third and the most viable route are to recycle the waste a few times previously it is released (Agrawal et al. 2015).

2 Textile Recycling and Reuse

Recycling implies the breakdown of a thing into its unrefined materials with the end goal that the rough material can be recuperated and used as a piece of new items. On the other hand, recycle insinuates a present thing being used again inside a comparable creation chain. Textile material recycling is the strategy by which old pieces of clothing and diverse materials are recovered for recycle or material recovery. It is the explanation behind the material recycling industry. Material recycling may

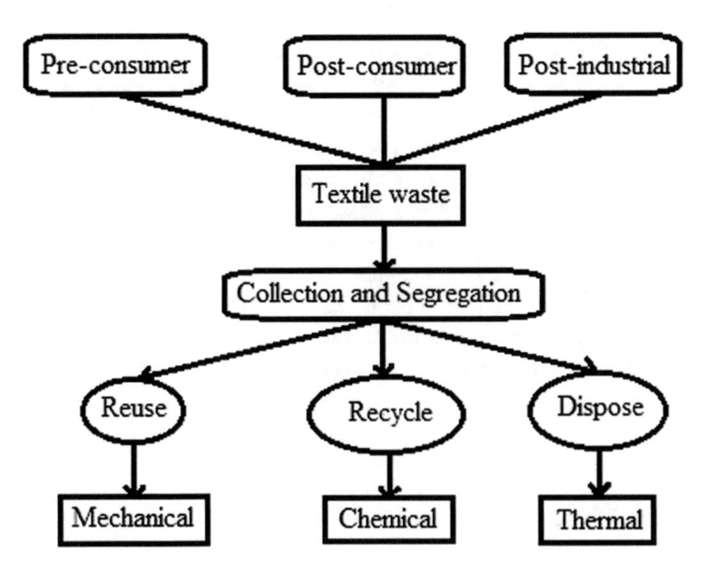

Fig. 1 Types of textile waste and recycling process

incorporate recouping pre-consumer waste or post-consumer misuse (Fig. 1). There are different ways to deal with perceive the sorts of recycling possible inside the material and dress sets (Wang et al. 2003).

Textile material recyling impacts numerous substances and contributes fundamentally to the social obligation of the present society. By recycling, organizations can understand bigger benefits since they stay away from accuses related of dumping in landfills; in the meantime, material recycling adds to positive attitude related with environmentalism, work for hardly employable workers, commitments to philanthropies and debacle help, and the development of utilized attire to regions of the world where modest garments is required. Since materials are almost 100% recyclable, nothing in the material and attire industry ought to be squandered. Fiber utilization development is a twofold edged sword in that while expanded fiber utilization stimulates the economy, it likewise contributes altogether to the issue of transfer. As shoppers keep on buying at a rate that meets needs as opposed to needs, the issue of what to do with squander is intensified. Material waste is made out of both common and engineered materials, for example, cotton, fleece, polyester, nylon and spandex (Yin et al. 2013). After engineered filaments went onto the market in the mid twentieth century, material recycling turned out to be more intricate for two unmistakable reasons: (1) expanded fiber quality made it harder to shred or "open" the fibres, and (2) fiber mixes made it harder to purge the arranging procedure. As of late, an assortment of procedures and innovations have been executed for recycling materials, including the Council for Textile Recycling (CTR) require a zero-squander objective by 2037. This hearty objective can be expert if the business and buyers grasp a comprehensive approach, shape key organizations, and uplift faithful utilization. Zero waste concentrates on a closed-loop mechanical/societal framework whereby squander is viewed as remaining crude material or an asset for esteem included items. This may include updating the two products and procedures keeping in mind the end goal to enable waste to be remanufactured into new items. Zero waste ideas consider the whole life cycle of products. The procedure of material recycling happens when post mechanical or post-consumer squander enters the recycling pipeline through assembling waste accumulation or passing on to someone else for reuse. Tragically, this procedure does not block material waste from winding up in landfills. More research is expected to precisely survey natural benefits of coordinating reused materials into new items. The sheer assorted variety of material kinds prompts difficulty in separating fibres for efficient reusing, with the vitality consumed to gather, sort and make the poor into another product convoluting the investigation (Muthu et al. 2012).

3 Types of Waste from Textile Industry

The waste created from material industry can be grouped into two sorts in light of the material use as

- Pre-consumer waste
- Post-consumer waste

3.1 Pre-consumer Waste

Pre-consumer waste is a material that was disposed of before it was prepared for customer utilize. Pre-consumer recycled materials can be separated and revamped into comparative or diverse materials, or can be sold as such to outsider purchasers who at that point utilize those materials for buyer items. Pre-consumer material waste for the most part alludes to squander results from fiber, yarn, material, and clothing fabricating. It can be process closes, scraps, clippings, or merchandise harmed amid creation, and most is recovered and recycled as crude materials for the car, furniture, sleeping cushion, coarse yarn, home outfitting, paper, and different ventures. Pre-consumer squanders are produced all through the first phases of the inventory network. In the crude materials area (fiber and yarn creation), ginning squanders, opening squanders, checking squanders, comber noils, brushed waste yarns, meandering squanders, ring-turning waste fibers, ring-spun squander yarns, open-end spinning waste fibers, and open-end spinning yarn squanders are usually gathered for recycling (Rosnev et al.).

3.2 Postconsumer Waste

Postconsumer material waste for the most part alludes to any item that the individual does not require anymore and chooses to dispose of because of wear or harm and regularly incorporates utilized or worn apparel, bed cloths, towels, and other buyer materials. Postconsumer squander which can be recuperated are dress, wraps/window ornaments, towels, sheets and covers, clean clothes and sewing leftovers, table materials belts totes matched shoes and socks. Post-consumer waste is brought from general society, which incorporates things that have no more use for the proprietor. This usually incorporates gave and disposed of clothing and some plastic things, for example, plastic jugs produced using polyethylene terephthalate. Nylon can likewise be recycled and extensive wellspring of post-consumer nylon squander is fishing nets left in the sea (Platt 1997).

4 Significance of Fibre Recycling

There are a few drawbacks related with the present land filling of textile fibrous waste. Initially, a tipping charge is required. Second, because of ecological concerns, there is expanding interest to forbid polymers from landfills. Third, land filling polymers is a misuse of vitality and materials. Assortments of advances have been created because of client requests for recycled products and as contrasting options to land filling. Aside from the instance of direct reuse, which is a typical type of usage for disposed of materials, some preparing is included to change over the waste into an item (Zamani 2011).

As the textile, attire and retail businesses move to wind up noticeably more practical, a region of intrigue is the utilization of recycled fiber, yarn, texture, and item content in the advancement and generation of new items. The choice to utilize reused materials in items must happen during plan and item improvement and proceed all through assembling forms. There are two phases in reusing—gathering and handling. Recycled products utilized as a part of material and clothing items can be acquired all through the material and attire store network and post-consumer accumulation techniques. The utilization of recycled crude materials lines up with the bigger developments of worldwide ventures toward a circular economy and closed-loop generation.

Textile material recycling is for both, natural and financial advantages. It maintains a strategic distance from many contaminating and vitality serious procedures that are utilized to make materials from crisp materials.

- Demand is diminished for material chemicals like colors and settling specialists.
- The prerequisite of landfill space is lessened. Materials prompt numerous issues in landfill. Engineered filaments don't disintegrate. Woollen articles of clothing do deteriorate yet create methane, which adds to an Earth-wide temperature boost.
- Leads to adjust of instalments as one purchase fewer materials for our prerequisites.

- As filaments get locally accessible, they don't need to be transported from abroad in this manner diminishing contamination and sparing vitality.
- Lesser vitality is expended while handling, as things don't need be re-colored or scoured.
- Waste water lessens as it doesn't need to be completely washed with huge volumes of water as it is improved the situation, say, crude fleece.
- Pressure on new assets too is diminished.

5 Characteristics of Recycled Fibres

Recycled fibres are appropriate to making nonwoven and yarns. When compared with initial primary fibres, recycled fibres demonstrate diverse qualities. The harm they endure at the time of production involves a wide range of fiber lengths with a high offer in short fibres and in addition strings and bits of texture not separated. Characteristics of recycled fibres are impacted by the loss being referred to, its pre-treatment and the separating procedure all things being considered. Usually, recycled fibres are accessible as mixes. Similarly as with the preparing of recycled fibres into yarns, nonwovens require extents of separated filaments to be as high as could be expected under the circumstances. Their lengths ought to be adequate to experience the spinning or web development process being referred to. Fragments of yarn or string still contained in the mix of recycled fibres directly add to lattice development in the nonwoven or they are additionally separated into filaments during the checking procedure. Recycled fibres are promoted at low value, essential fiber materials made of regular or engineered substances, adding to holding costs down. When compared with initial primary fibres, the nature of recycled filaments is difficult to characterize. The estimating procedures and gear routinely utilized are not extremely supportive here. This is because of the mixes of recycled fibres being non-similar and rich in short filaments and in textile material remains, which are not fibre-type (Tamer and Mohamed 2014).

6 Recycling Process

Recycling is the breakdown of an item into its crude materials. For quite a long time, textile products such as fabrics and garments were separated to the yarn arrange and the yarn was utilized to deliver distinctive weaved or woven textures. Now and again, the yarns are additionally separated to the fiber stage and afterward the fibers were respun into yarns to be utilized as a part of new material items. Inclinations for recycling of material squanders in the business appear to the overwhelmingly thermoplastic polymer-based fibers because of the ease and plausibility of reprocessing them. Besides, these materials can go up against various structures and shapes subsequent to recycling. Regular natural fibers, for example, cotton, fleece, and silk are likewise finding their routes into recycling streams.

The two essential stages engaged with recycling are gathering and reprocessing. With respect to textile materials and attire industry, the accumulation procedure happens at different focuses through the inventory network, and there are programs where the general population can be associated with the procedure. Squander is additionally gathered from sources outside the material and clothing industry for reprocessing and use in attire and material finished results. Reprocessing of the gathered materials is basic in deciding if it will add to an open-or closed circle framework. There are various explanations behind barring material from the circle. Two basic purposes behind this prohibition are: (1) debasement of the crude material that outcome in decreased quality and (2) consolidation of the crude material into an item that isn't recyclable.

6.1 Open-Loop Recycling (OLR)

Open-loop recycling alludes to a framework in which an item's crude material is separated to be utilized as a part of a moment, frequently random item framework. For the most part, the second item won't be recycled and rather be discarded toward the finish of its life. Thus, OLR is a commendable recycle of the principal item's virgin materials and lessens the need to expend virgin materials in the fabricate of the second item. In any case, OLR ordinarily just postpones a material's definitive entry to municipal solid waste, as there are cut-off points to how regularly a material can be recycled without quality being corrupted (Fig. 2). Ordinarily, OLR recycled things in textile and attire incorporate

1. Pre-consumer textile material waste, for example, offcuts from the cutting procedure;
2. Post-consumer material waste as entire pieces of clothing; and
3. Post-consumer PET containers that might be fabricated into recycled PET (RPET) fiber.

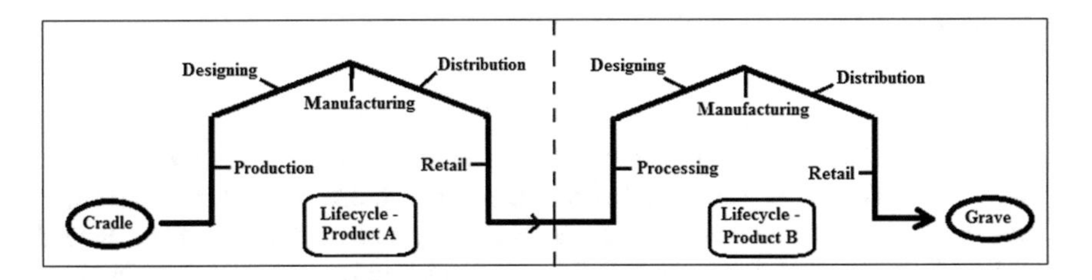

Fig. 2 Open-loop recycling (Payne 2015)

OLR has demonstrated achievable in the fashion setting, both in accumulation of pre-and post consumer material waste for use in different items, and in gathering of utilized jugs or bottles for recycling into materials. As of now, OLR of PET bottles or jugs to fibre has had the best accomplishment for recycle as a material in the textile and fashion area, with an open circle of waste from the first item (PET jugs) utilized as feedstock for the second item framework (polyester texture to piece of clothing). A typical approach is to mix the recycled yarns with virgin fibres to make materials that are of clothing quality (Curran 2012).

Conversion of PET bottle to fibre using OLR

PET bottles or jugs to fibre is an OLR strategy in which PET containers are recycled into PET flakes, re-spun into fibre and afterward woven or sewed into materials. Not at all like the technique portrayed above, where material waste enters a moment product life cycle, this OLR approach sees squander from other item cycles being used in material and clothing creation. As PET is non biodegradable, recycling is a proper utilization of the asset. PET containers have been recycled since the 1970s. Be that as it may, the nature of the subsequent recycled polyethylene terephthalate (RPET) relies upon whether the bottles or jugs to be recycled contain contaminants including dust, shading, acids and water. The procedure for mechanically recycling PET jugs to fibre is as per the following:

1. Containers are gathered, cleaned and arranged by shading.
2. Marks are expelled.
3. Bottles or Jugs are handled into PET flakes.
4. The PET flakes are liquefied and expelled through the spinneret into new fibres.

The nature and quality of RPET in mechanical preparing is for the most part not high when contrasted with virgin PET as virtue depends intensely on polluting influences or contaminants inside the containers. And additionally the mechanical procedure, PET can be recycled utilizing a substance procedure. Stages 1 to 3 are indistinguishable to the above; the PET flakes are chemically changed and come back to an oligomer or monomer.

6.2 Closed-Loop Recycling (CLR)

CLR alludes to recycling techniques whereby the material being recycled is a similar material being delivered: 'an item enters the generation chain of a similar item again after utilize'. The waste material or fiber re-enters a piece of clothing generation chain, both pre-and post-consumer mechanically recycled materials might be viewed as closed-loop recycled. Successful CLR has demonstrated more troublesome at scale than OLR. Figure 3 represents the closed-loop lifecycle of textile material.

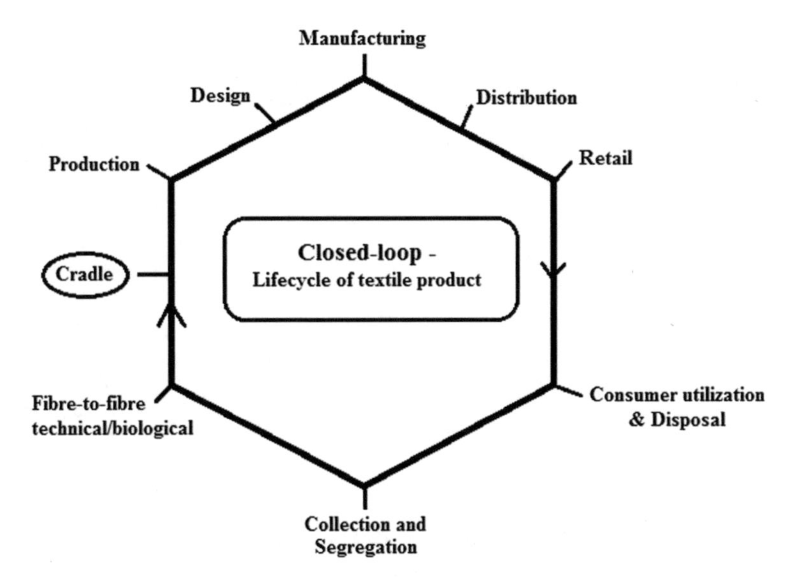

Fig. 3 Closed-loop recycling (Payne 2015)

Support to-support CLR (Cradle-to-cradle)

The support to-support philosophy is a radical way to deal with CLR in which a CLR fiber will be recyclable and also reused into a similar creation chain. C2C closed-loop framework, squander is recovered and utilized again in the generation of results of the same or higher esteem. Squander is redirected into either organic or specialized streams. Organic waste can be treated with the soil, while specialized waste can be recycled inside industry to make similar items once more. A related closed-loop approach is in the recycle of articles of clothing. Both recycling of garments and furthermore upcycling of apparel are identified with CLR. Closed-loop reuse suggests apparel can have different valuable lives on the second-hand showcase. In spite of the fact that recycle of articles of clothing isn't recycling in the feeling of separating an item into its crude materials, it compares to CLR in that the item may enter another life cycle inside a similar generation chain (Payne 2015).

7 Textile Recycling Approaches

In the clothing industry, primary or essential recycling is the social event of pre-consumer offcuts of the surface from creation. Cut-and-sew manufacture of dress infers that there is basic material waste happening on account of the additional surface of solitary pattern material.

Secondary or Auxiliary recycling incorporates aggregation and reusing of post-consumer material misuses, for instance, pieces of clothing, materials identified with

Table 1 Stages in textile recycling (Wang 2010)

S. No	Stages	Process
1	Primary	Recycling materials from industries
2	Secondary	Conversion of post-consumer product into useful material
3	Tertiary	Conversion of plastic wastes into fuels
4	Quaternary	Burning waste as a method of recovering the fixed energy

families, and so forth. Dependent upon consumer action, this clothing may be sent to civil strong waste or provided for philanthropies or social event organizations. Garments gathering association will then sort the dress into higher quality articles of clothing proper for resale and lower quality clothing sensible for recycling.

Tertiary reusing incorporates the engineered or chemical degradation of nylon or polyethylene terephthalate (PET) for repolymerisation. This requires either spotless, organized pre-consumer waste or post-consumer misuse, orchestrated by nature of fibre, as a feedstock.

Quaternary reusing insinuates devouring the fibrous solid strong waste and utilizing the heat vitality delivered. In quaternary reusing, the embedded essentialness can be recuperated through consuming procedure or burning. Overall recycling stages and its process ia represented in Table 1.

8 Methods of Recycling Fibres

8.1 Mechanical Recycling

Mechanical recycling procedures can bring about the creation of texture, yarns, or fibers to be utilized as a part of new textile products. The disposed of material is opened up, clothing is dismantled, and textures are cut into little pieces. It is then gone through a spinning drum to proceed with the breakdown and fibers are acquired. This procedure is known as garneting. The subsequent fiber attributes of length, fineness, quality, polymer, and shading decide the quality and what the most suitable new final result would be. Ordinarily, squander gathered from the assembling store network will create higher-quality recycled fibers that those gathered from post-consumer squander. The pre-consumer and post-industrial handled waste can be respun into yarns which woven or weaved into textures, and after that utilized as a part of attire,

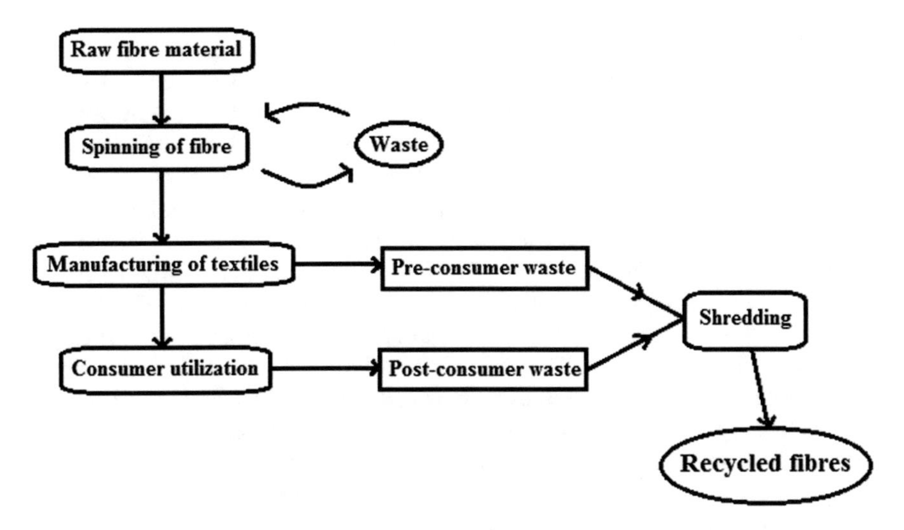

Fig. 4 Mechanical recycling of fibres

sheeting, and upholstery. Mid-range review fibers can be utilized to make textures however are utilized as a part of final results, for example, wipes and fillings. Lower-quality fibers will be utilized as fortification in different structures (i.e., concrete), nonwoven textures, cover underlay, shoe decorates, car sound and warm protection, home protection, stuffing for toys, and opposite final results. Plastics, including plastic containers and thermoplastic fibers, are generally recycled utilizing mechanical strategies. In these cases, the plastic waste is cleaved into little flakes that are liquefied and after that expelled into a shape to be utilized as a part of another item. This liquefy can be expelled into filaments, yarns, or other framed items. There is a little discernible contrast between virgin polyester and recycled polyester fibers. This is a typical strategy for reprocessing the plastic water bottles and fishing nets. In any case, not all recycled thermoplastic fibers have properties like virgin fibers (Hawley 2006). The outline of mechanical recycling process is given in Fig. 4.

8.2 Chemical Recycling

Chemical recycling is the other technique regularly used to process the gathered waste in the material business. Manufactured fibers including polyesters, polyamides, and polyolefins can be artificially reused. This falls under the tertiary class of recycling which requires the separating of the manufactured fibers for repolymerization. This procedure can be utilized when PET plastic water bottles are reused. Regardless of whether it is the accumulation of utilized polyester attire, texture scraps, yarns squander, or different plastics, the recycled things are broken into little pieces from which chips are delivered. The steps involved in chemical recycling are presented in Fig. 5. The chips are deteriorated to frame dimethyl terephthalate, which is then repolymerized and spun into new polyester fibers, filaments, and yarns. Mixes are specifically testing to recycle due to the different physical and chemical properties of the fibers in the waste. Cotton and polyester mixes are a standout amongst the most regularly utilized attire and home material things. Chemical recycling has demonstrated effective when utilized with mixed materials as it utilizes a specific corruption strategy. In results of cotton and polyester, the fibers can be artificially isolated and afterward transformed into new fibers. Right now, there is a procedure being created utilizing n-methyl morpholine-N-Oxide, which breaks down cellulose. The broke up cellulose and polyester are isolated by filtration and the caught polyester is respun into a fiber, filament, or yarn. The broke down cellulose can be utilized as a part of the generation of recovered cellulosic fibers including Lyocell. Nylon and spandex is a mix usually found in superior sportswear and dynamic wear. Usually, the level of nylon is considerably more noteworthy than that of spandex and nylon can be recycled and reused. It is realized that spandex can be expelled from mixed textures by dissolving it in solvents, for example, N,N-dimethyl formamide. Nevertheless, this dissolvable is costly and there are environmental issues with its utilization. There has been accomplishment by first treating the mixed texture with warmth to corrupt the spandex, and after that presenting the texture to a washing procedure utilizing ethanol, which successfully evacuates the spandex deposit leaving just the nylon.

Today, for results of single fiber content textures, mechanical recycling is more noticeable. The substance recycling systems require more vitality utilization and there is high capital speculation so this choice is common for substantial scale producers. As the innovation enhances, the interest for recycled content increments and

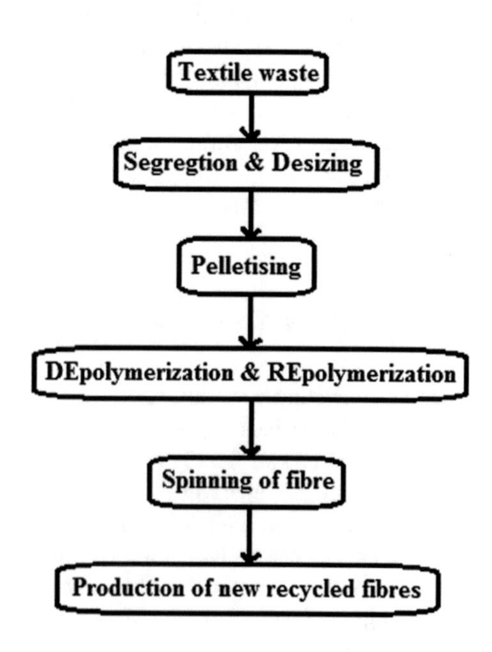

Fig. 5 Chemical recycling of fibres

as the cost of virgin crude materials increments, there is probably going to be a move from mechanical to chemical recycling of these materials (Dang et al. 2002).

9 Applications of Recycled Fibres

Recycled fibres can be made from a diversity of textile waste. Both quality and process capacities of such fibres rely upon the sort of waste. Conventional are the pure arranged fibres of astounding which are accomplished from spinning fibre waste. Interestingly, recycled fibres produced using end-of-life materials are of considerably poorer quality. They will seldom be found of homogenous fibre compose. There are numerous routes open to utilizing recycled fibres in both textile material and non-textile material items. The acceptability of procedures relies upon squander attributes and on the amount they cost.

9.1 Yarns

The waste created in a textile plant is a vital factor in deciding the working expense and in this way in affecting mill benefits. The recovered fibres from waste can be utilized to deliver mixed yarns in various parts. These fibres can be reused for the open end spinning and contact spinning however present process endeavours on ring spinning are likewise in advance. Results likewise showed that the recovered fibres have a decent clean capacity which permits its mix with virgin filaments. This yarn can be woven or weaved for some uncommon reason yet till a date it can't satisfy crafted by virgin filaments. Scientists guaranteed that yarn can be utilized for particular utilize if squander is chosen in particular sums from the different waste classifications and combined legitimately. Endeavour on making Dref yarns from reused strands likewise demonstrates great outcomes however as crude materials utilized is thoroughly squander materials so textures created obviously has a place with shabby textures which are recommended to be utilized as a part of the field of cleaning fabrics, wrapping fabrics and covering fabrics (Agarwal et al. 2015).

9.2 Biocomposite and Nonwoven Materials

Protein fiber squanders, for example, by-products from the fleece material industry, low quality crude fleeces not fit for spinning, speak to a vital sustainable wellspring of biopolymers. Fleece fibres when damaged in their histological parts connecting the intercellular bond by ultrasonic-catalyst medications, the subsequent cells were implanted in a polymeric film-shaping network of cellulose acetic acid derivation, of acquiring new composite material, appropriate for film creation and fiber spinning.

Cellulose acetic acid derivation is to a great extent utilized as a part of the generation of yarns for materials, channels, plastics; electrical protections, photographic movies, pigmented sheets, therapeutic and clean application since it display an exceptional protection from molds and microorganisms. Plastics and textile fibres with novel properties, for example, enhanced imperviousness to fire, dampness recover, coloring exhibitions and shading impacts, handle and look may be created from new composite materials with combined properties of man-made and protein polymers which are normally hydrophilic, non-consuming and colored well by the greater part of the business color stuffs. Recycled fibres can be viewed as traditional in specialized materials, especially in nonwoven. Recycled fibres are utilized as a result of low costs or in light of the fact that they only cover something up. In any case, recycled fibres are additionally connected in nonwoven to use exceedingly profitable practical segments.

9.3 Commercial Recycled Textile Products

Cotton and polyester are presumably the most well-known fibers recycled, yet different fibers including fleece, nylon, and even aramids are being recycled in yarn generation. There are many yarn producers joining recycled content into their items. The accompanying are a few yarn makers and their business items.

Repreve

Repreve is a brand of recycled fiber that are produced using recycled polyester including post-consumer plastic jugs and post-industrial waste from assembling squanders. Utilizing post-customer squander balances the need to utilize new assets and in this manner, there is a lessening in the generation of nursery gasses. It is delivered by Unifi.

Ecocircle

Ecocircle is a fiber created from recycled polyester. The procedure is a fiber-to-fiber polyester reusing framework that was produced by Teijin Fibers. It is a closed-loop recycling framework for polyester items and a chemical recycling procedure is utilized. The textures delivered from Ecocircle are creative and produced for use in the pieces of clothing.

Nylon 6

Nylon 6 can likewise be recycled and the Econyl Regeneration System was presented in 2011. The Nylon 6 polymers are delivered utilizing both post-consumer squander and pre-consumer squander. A significant part of the post-consumer squander originates from fishing nets disposed of in the sea and the heap of utilized cover. They keep on increasing the waste accumulation system and gather materials for recycling all through the world.

Patagonia

Patagonia has a few built up recycling programs and effectively utilizes recycled fiber in their items. The recycled programs are set up utilized for polyester, fleece, and cotton. They began utilizing fiber-to-fiber reusing framework to keep utilized attire items out of the waste stream and junk incinerators. They additionally gather post-industrial assembling waste and post-consumer exhausted articles of clothing for reprocessing and use in new attire. Patagonia likewise utilizes recycled fleece in its fleece items. The recycled fiber is joined with virgin natural cotton (Patagonia 2016).

Ecosmart

Hanes presented another line, EcoSmart in 2010. This line incorporates clothing things with recycled cotton or potentially polyester fiber content.

10 Benefits of Textile Recycling

- Recycled garments lessen the landfill space. Landfill destinations represent a risk to the earth and water supplies. When it downpours, water depletes through the disposed of garments and grabs perilous chemicals and fades. This water ends up being lethal. Material produced using engineered fibres won't deteriorate rapidly though textures like fleece discharges methane, during disintegration and the two fibres eventually cause a worldwide temperature alteration. At the point when these textures are recycled, this risk will be diminished to a significant degree.
- It saves money on utilization of vitality, as recycled garments require not be re-colored or sourced. Lessened utilization of colors and chemicals limits their make and eventually the unfavourable impacts of their fabricate.
- It decreases cost of acquiring new materials and expands profit efficiency.
- It additionally limits the expenses of transfer of new virgin or crude materials and furthermore the ecological effects by decreasing utilization of new crude materials and delivering items from prior one.
- Textile recycling does not make any new unsafe waste.

11 Challenges of Textile Recycling

- There is no economical motivating force for waste makers to lessen squander.
- Low esteems, high transportation cost or absence of market interest for recouped materials especially.
- The power of little and medium recuperation and recycling undertakings debilitate interests in squander recuperation advancements.

12 Future of Recycled Textile Materials

Innovations in technology are being continuously investigated in the field of textile material recycling. A key hindrance to powerful recycling is the multiplication of textile materials in various fiber mixes that are hard to isolate for recycling, for example, cotton and polyester. Analysts have inspected how to isolate the cotton from the polyester utilizing earth sound ways to deal with breakdown of cotton so as to recover the polyester for recycling. Another boundary to successful recycling is the subsequent poor shading nature of low-grade. The dim dark grey of recycled fiber is unsuited to most attire applications or for sure for most broad items. A Japanese report proposes a shading coding to empower the subsequent low-grade to be all the more effectively handled into usable yarns. As both mechanical and chemical recycling innovations grow, more open doors may show up for CLR of materials inside attire, and also proceeded with OLR of materials into different items. Patterns towards natural concerns see shoppers more prone to anticipate that organizations will offer economical materials and think about the biological effects of their items. Extensive retailers have the compass to gather utilized apparel at scale, and by taking a lead they show that they see the social incentive and additionally financial incentive in seeking after recycling alternatives (De Silva et al. 2014).

13 Conclusion

Numerous open doors exist for enhanced textile material recycling. Examination of the vitality and water use favourable circumstances in using recycled materials is obvious crosswise over LCAs of both OLR and CLR fibres. Anyway, impressive obstructions exist towards more prominent take-up of recycled fibre with respect to organizations. Furthermore, in spite of the fact that there have been progresses in the previous years, more efficient recycling and gathering of textile material waste is required from various partners, for example, attire producers, people and governments. The favourable circumstances for utilizing recycled fibres inside articles of clothing are quantifiable in LCA crosswise over water, vitality and land utilize pointers. Numerous partners can assume a part in enhancing textile recycling rates and recyclability. Originators and producers can assume a significant part in reverse and forward building pieces of clothing to be dismantled for simplicity of recycling and furthermore in choosing materials that contain a level of recycled fibre. Purchaser patterns towards all the more biologically considered items are as of now inciting the attire part to effectively use recycled yarns. A lot of material waste is discarded in landfills every year. That not just stances financial and ecological issue to the general public yet additionally speaks to a serious misuse of assets. Despite the fact that the natural consciousness of the overall population has expanded essentially as of late, still their ability to effectively take an interest in reducing waste by recycling should be improved. Fiber recycling advances, use and scope of use of recycled fibres will turn into a convenient device to acclimatize the loss as esteem included product. To enhance recycling, all encompassing thoughtfulness regarding the store network must be actualized. The ecological impression can be diminished if both science-based and societal based methodologies over the textile material production network are considered.

References

Agrawal Y, Kapoor R, Malik T, Raghuwanshi V (2015) Recycling of plastic bottles into yarn & fabric. Available at http://www.textilevaluechain.com/index.php/article/technical/item/247-recycling-of-plastic-bottles-into-yarn-fabric

Bhatia D, Sharma A, Malhotra U (2014) Recycled fibres: an overview. Int J Fibre Text Res 4(4):77–82

Curran MA (2012) Life cycle assessment handbook: a guide for environmentally sustainable products. Wiley, Hoboken

Dang W, Kubouchi M, Yamamoto S, Sembokuya H, Tsuda K (2002) An approach to chemical recycling of epoxy resin cured with amine using nitric acid. Polymer 43:2953–2958

De Silva R, Wang X, Byrne N (2014) Recycling textiles: the use of ionic liquids in the separation of cotton polyester blends. RSC Adv 4:29094–29098. https://doi.org/10.1039/c4ra04306e

Hawley JM (2006) Digging for diamonds: a conceptual framework for understanding reclaimed textile products. Clothing and Textiles Res J 24(3):262–275

Muthu SSK, Li Y, Hu J-Y, Ze L (2012) Carbon footprint reduction in the textile process chain: recycling of textile materials. Fibers and Polymers 13(8):1065–1070

Patagonia (2016) Recycled wool [Online]. Available from http://www.patagonia.com/us/patagonia.go?assetid=93863

Payne A (2015) Open and closed-loop recycling of textile and apparel products. In: Handbook of life cycle assessment (LCA) of textiles and clothing, pp 103–123

Platt (1997) Weaving textile reuse into waste reduction. Institute for Local Self-Reliance, Washington

Roznev A, Puzakova E, Akpedeye F, Sillstén I, Dele O, Ilori O. Recycling in textiles. HAMK University of Applied Sciences Supply Chain Management

Tamer FK, Mohamed ED (2014) Recycling of textiles. J Text Sci Eng S2:001. https://doi.org/10.4172/2165-8064.S2-001

Wang Y (2010) Fiber and textile waste utilization. Waste Biomass Valorization 1(1):135–143

Wang Y, Zhang Y, Polk M, Kumar S, Muzzy J (2003) 16—Recycling of carpet and textile fibres. Plastics and the environment: a handbook. Wiley, New York, pp 697–725

Yin Y, Yao D, Wang C, Wang Y (2013) Removal of spandex from nylon/spandex blended fabrics by selective polymer degradation. Text Res J 84(1):16–27

Zamani B (2011) Carbon footprint and energy use of textile recycling techniques. M.S. Thesis, Chalmers University of Technology

Sustainable Apparel Production from Recycled Fabric Waste

R. Rathinamoorthy

Abstract Waste generation in the apparel industry is one of the unavoidable factors of the garment production. Cost saved in cutting (fabric) is the cost saved in the overall production of the garment. This is because; the cutting department decides the amount of fabrics utilized for production and for waste. Fabric approximately covers half of the garment cost, and is the major contributor to garment cost. This research work aims to validate the enterprise benefits by utilizing this waste fabric for the reproduction of the apparel fabric. As the world moves towards more an eco-friendly and sustainable production, the fashion and apparel industry also in the path, by producing organic material. This research is focused on recycling the waste fabric into fibers. The main aim is to develop new apparel product by using recycled fiber, yarn, fabrics. For this purpose, the cutting waste from the knitted garment manufacturing industries was collected and utilized for the recycling process. It was found that the fabrics had a reasonable amount of physical characteristics like dimensional stability, bursting strength, pilling and abrasion resistance. The fabrics were made into casual garments and cost-effectiveness study also performed on the developed garments. The results were promising that the developed garments from recycled fibers are cheaper than the garments developed from normal raw materials. It is also double-advantageous for the manufacturers by providing some earning out of the waste and also solves waste management and disposal issues.

Keywords Fast fashion · Fabric waste · Recycling · Physical properties
Cost benefit analysis · Sustainable fashion brands

1 Introduction

Fast fashion is a term used to describe clothing collections that are based on the most recent fashion trends. The fast fashion collections are generally adaption from

R. Rathinamoorthy (✉)
Department of Fashion Technology, PSG College of Technology, Coimbatore 641004, India
e-mail: r.rathinamoorthy@gmail.com

S. S. Muthu (ed.), *Sustainable Innovations in Recycled Textiles*, Textile Science and Clothing Technology, https://doi.org/10.1007/978-981-10-8515-4_2

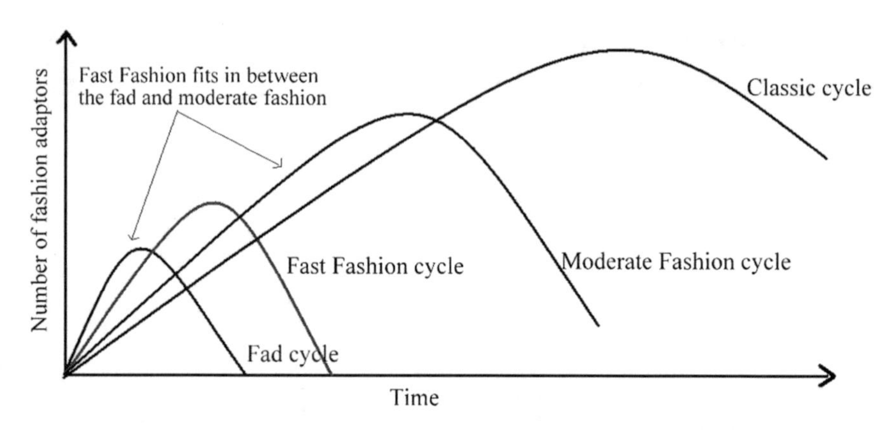

Fig. 1 Comparison of the acceptance cycle of fad, fast fashion, moderate fashion and classics

current high fashion luxury trend. The main advantage of the fast fashion system is its very nature, a fast-response system and that encourages disposability. This is a concept dominated by consumption, fast-changing trends, and low quality; leading consumers buy more clothes because they are affordable but discard these after only one season (Fletcher 2008). The fast fashion companies thrive on fast prototyping, numerous designs, rapid delivery and also the delivered materials are floor ready with a price tag to sell (Skov 2002), in contrary to the old turnaround time, which will usually take six months, from the catwalk to consumer (Tokatli and Kizilgun 2009). The following diagram represents the life cycle of an apparel product. In earlier times product life cycle was mentioned as a classic which typically consists of the introduction, acceptance, culmination, and decline stages. But the fast fashion products fall in between the fad and moderate fashions as in Fig. 1. Where fad is a short-lived fashion that suddenly becomes popular and quickly disappears and the moderate is prevalent in a moderate period, there is a big possibility for it to be cyclical (Solomon and Rabolt 2004).

In general the fast fashion aimed to reduce the process and lead time in the buying cycle and helps in getting the new product to the customer as soon as possible to satisfy the needs while the trend in peak (Liz and Gaynor 2006). The fast fashion cycle made the companies to work on up to the minute design and made the customers to refresh their wardrobes very often. Since, the consumer spending increased by fast fashion, the consequences increases the production which in turn maximizes the impact on the environment. A study conducted by McKinsey & company revealed that the environmental impact of the clothing manufacturing companies will increase 80% in 2025, if the fast fashion trend continues and the world market consumes to the same western per capita level consumption (Remy et al. 2017). The Fig. 2 shows the environmental footprint of the apparel industry in terms of land, water and CO_2 emission (Remy et al. 2017).

As an impact of the fast fashion trends, the leading apparel brands also changed their conventional ways of presenting garments in for six month season. Currently,

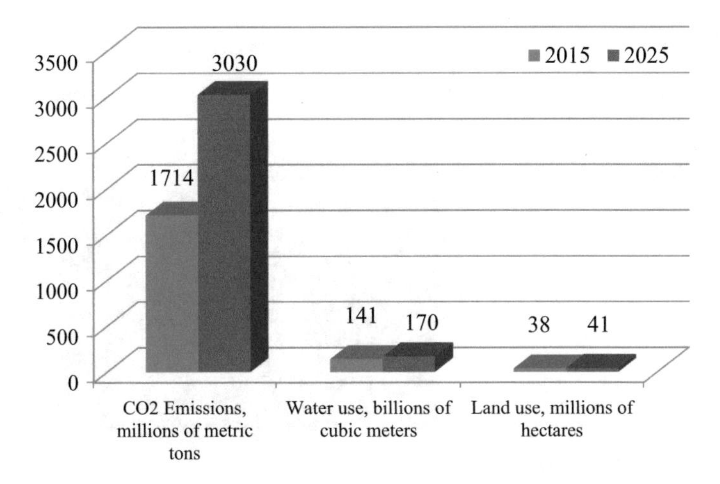

Fig. 2 Environmental impact of the clothing manufacturing companies at 2025, estimated based on current per capita consumption from developed countries (*Source* www.mckinsey.com)

24 new collections were offered by Zara for every year and 12–16 collections offered by H&M (Remy et al. 2017). Among all European apparel companies, the average number of clothing collections has more than doubled, from two a year in 2000 to about five a year in 2011. Due to the quick response and low cost fresh designs, the European fast fashion market grown faster than the retail market along with higher profit margin of 16% than the traditional retail shop, which has an average of 7% (Sull and Turconi 2008). As a consequence of the fast fashion strategy, the manufacturing industries and brands started to produce more and so contributing significantly increased the level of environmental impact (Modi 2013).

2 Textile and Apparel Industry–Current Indian Scenario

The Indian textile industry is currently estimated around US$ 108 billion and at 2021, the industry is expected to reach US$ 223 billion. Next to agriculture, textile and apparel industry provides over 45 million direct employments and approximately 60 million indirect employments (Shekhar 2017). Approximately 5% of India's total Gross Domestic Product (GDP) was contributed by the Indian textile Industry. At the same time, exports are implied to increase to US$ 185 billion from approximately US$ 41 billion currently (Shekhar 2017). The Indian textile industry is the second largest manufacturer and exporter in the word next to china. India has a share of 5% of the global trade in textile and apparel. Out of the total Indian exports 15% share belongs to textile and apparel in year 2015–16, compared to 13.6% in the year of 2014–15. The total textile and apparel export during 2016–17 (up to April 2017) is

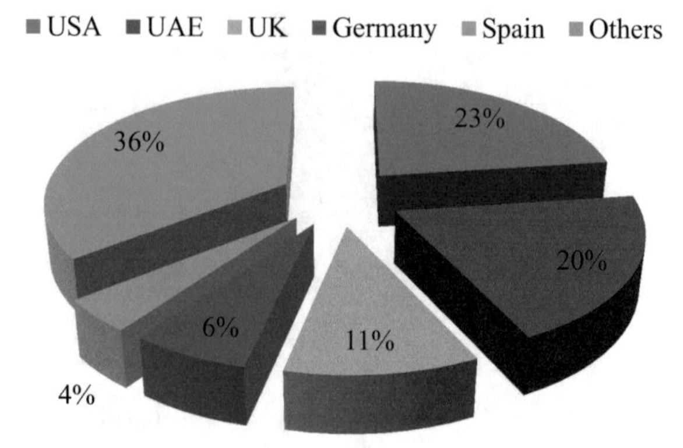

Fig. 3 Percentage share of various countries to India's total apparel exports for the financial year 2016 (Modi 2016) (*Source* www.careratings.com)

valued 18.7 billion USD, with a share of 14% in India's total export of 132 billion USD during the same period (Annual Report 2016).

Out of the textile sector, readymade garment or apparel manufacturing is the largest segment in the Indian textile sector. It accounts for 60–65% of the total textile industries (Modi 2016). The apparel sector is also acts as a largest source of foreign exchange flow in the country. As per Annual report 2015–16, by Ministry of Textile, India ranked as the 6th largest exporters in the world after china, Bangladesh, Vietnam, Germany and Italy. In 2015–16, it is noticed that, the apparel trade decreased around 5% due to various reasons including weak demand from European countries and Japan (Annual Report 2016). Despite the weak global apparel trade, the Indian apparel exports grew by 1% in the financial year 2016 than 2015 (Modi 2016). Figure 3 represents the percentage share of various countries to India's total apparel exports for the financial Year 2016.

As per the Apparel export and Promotion Council (AEPC) 2016–17 annual report, the total export in India was hiked a growth of 5.4%. In rupee terms, export for the period of 2016–17 was Rs. 1,17,202.4 Cr. as against Rs. 1,11,182.8 Cr. of the same period of previous financial year. Out of that total export, the knitted export was calculated as 8267 million USD (2016–17), with the share of 47.3% and a Growth of 7.9%. The woven Exports around 9212.0 million USD (2016–17), with a Share 52.7%, negative growth rate of (Decline) 1.2%. The Fig. 4 mentions the overall knitted and woven garment export for the previous year's (Apparel export promotion council-Annual report 2016).

The increase in the apparel product manufacturing is not only by the impact of the fast fashion trend but also by government policies and norms. The industry has witnessed a spurt in investment during the last five years. During the period of 2000 to march 2017, the industry attracted worth of US$ 2.47 billion as Foreign Direct Investment. Numerous export promotion policies also implemented for textiles sector by the Indian government. It has also allowed 100 percent FDI in the Indian textiles sector under the automatic route. Hence, altogether the government initia-

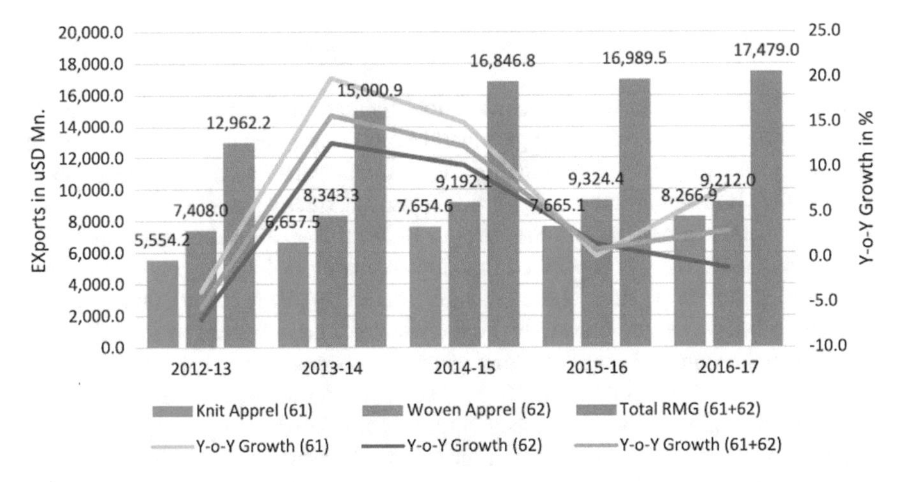

Fig. 4 Overall knitted and woven garment export for the previous year's along with year on year growth percentage (Apparel export promotion council-Annual report 2016) (*Source* http://aepcindia.com/)

tive and customer needs in the textile and apparel sector decently generates lots of requirements towards the fashion brands and manufacturers (Ministry of Textiles 2017).

3 Impact on Sustainability

Any apparel or textile material is defined as sustainable one, when it provides the maximum benefit to the people and minimum impact on environment in all means (Joy et al. 2012). In a basic term, every last piece of clothing has a environmental footprint at each phase in its creation. That is the reason there is an innate logical inconsistency between the fast fashion business model and the idea of environmental sustainability (Lejeune 2016). Fast-fashion industry portrays society's enthusiasm towards with industrious utilization. Thus it makes unsustainable demand, meaning that, this demand pushes the manufacturers even the big brands, to cut expenses and to deliver huge quantities at low prices. These prerequisites are clearly conceivable with the courtesy of cheap synthetic fibers, chemicals and dyes. Which in turn result higher environmental impact (Jake Hall 2016).

The impact of apparel industry ranges from water consumption in the case of cotton production, chemical pollution in the case of synthetic production along with high level of energy usage. Generally with the fast fashion, the use of synthetic fiber increased significantly to produce cheap products. It was identified that the polyester the most commonly used synthetic fiber which emits almost three times more carbon dioxide in its lifecycle than cotton. More over it can take decades to degrade completely. In year 2016 alone about 21.3 million tons of polyester used for clothing purpose, this is around 157% higher than that of the year 2000 (Cheeseman 2016).

In 2010, the garment industry association planned to improve the resource utilization across the lifecycle of the garment. In this regard they conducted a resource flow study in Tirupur by emphasizing the garment waste as a prominent source of potential. Before and after this till 2017, no constructive studies were made on the knit clusters in and around the Tirupur district. Based on the waste composition studies performed, it is notified that approximately 4% of the total waste used in landfill is textile wastes. These statistics are an aggregate of all sectors in the garment industry (i.e. pre-consumer, post-consumer and industrial waste). The data revealed that around 10.8 million tons of textile wastes were generated in only in the year of 2010 (Sakthivel et al. 2012). The recycling industries prevent only 1.5 billion pounds of textile waste by recycling. It is also informed by Tirupur Environmental Protection Agency that even after selling the waste cloths and fibers to the textile recyclers, including used clothing dealers and exporters, wiping rag graders, and fibre recyclers, approximately 400 million pounds of textiles collected by this agency (Sakthivel et al. 2012).

4 Waste Generation in Apparel Industry

4.1 Types of Wastes

In textile and apparel industries, there are different types of wastes are obtained and they are classified as three types namely

i. Production waste
ii. Pre consumer waste and
iii. Post consumer waste

Production waste—These are all the leftover items in the apparel manufacturing firm such as trimmings, proofs, leftover fabric, off-cuts, ends of rolls, etc. Production wastes are one of the important waste types due to its virginity and also used for many reuse, recycle and up cycling process. This is mainly because of the volume produced is generally quite large and regular.

Pre consumer waste—It is a material that was discarded before it was ready for consumer use. For example, in Tirupur, sometimes the over-estimated fabric meters and off-cuts of saleable size has been resold into markets or made-up into smaller items. Most of the time, most pre consumer textile wastes in Tirupur is simply sent to landfill. New York Times journalist Jim Dwyer published a story in January 2010, wherein he discovered several instances where the fast fashion retailers had asked their employees to cut holes in unsold garments and discard them. Competition for the biweekly changing collections becomes the basis of most of pre consumer waste generation in the apparel industry. The journalist also mentioned many alternative methods followed by various industries across the world (Dwyer 2016).

Post consumer waste—This types of waste consists any type of apparels or home textiles that an individual does not require anymore and chooses to dispose of due to different reasons like, worn out conditions or outgrown or out of fashion or etc., These kind of materials are normally of good quality, that can be recuperated and along these lines reused by another individual as second-hand garments a lot of which is sold to underdeveloped countries.

5 Waste Management Methods

The general waste management concept is 4 Rs.—reduce, reuse, recycle and recover. This concept decides the methods according to their desirability. This waste management concept insists, waste generation should be prevented or reduced to the possible extent as a first step. On the off chance that at all it is created, the waste ought to be reused in view of their probability. If not or after reuse, the waste can be recycled based on the material. Finally, if it is possible, the recovery options also can be tried to recover its raw material. In apparel industry, the purpose of the waste management system is to separate the most extreme practical benefits from garments while at the same time creating the base measure of waste and causing the minimum environmental impact. The following content discusses the 4R Principle in terms of Apparel and Fashion Industry.

Reduce—Meaning in general as less buying and using less amount of product in all possible way. The concept of zero waste in apparel and fashion industries is not a new concept. However, due to business motivations and other requirements the zero waste concepts were not implemented even at the possible situation. The kind of production, with no waste can be classified under reduce strategy. In some extreme cases, the reduce strategy can also implemented in rework. Repairing and reconditioning the apparel products as an acceptable one is also an initiative towards the waste reduction. The "reduce" concept should be kept in mind during all industrial manufacturing situations. This will help us to achieve possible reduction in the waste generation.

Reuse—It is an important strategy for apparel industry. It focuses on the reuse of rejected material, re-distribution and resale of rejected apparel product. In the case of apparel manufacturing industry, the reuse concept advices to utilize the waste fabric or material into a value added item for another product. The fabric from lay end wastes and splicing wastes can be used as a component or material for sampling process. In similar way the trims, accessories and other material can also be utilized. The rejected garments from factory can be sold in discounted price or as a second hand product in the market instead of being discarded or incinerated. This is applicable only if the reason for the rejection is minor or quality parameter issues like, stripes or check mismatch, pocket improper alignment or any other sewing defects. In the retailer or brand point of view, they can sell their backlog or old season's stock either in the second hand or in a discounted price. Reusing garments significantly helps reduce the negative environmental impact of fashion. For example, the energy used to collect, sort, and resell second-hand garments is between 10 and 20 times less than that needed to make a new item (Fletcher 2008).

In the other category, as an individual, every individual has their own unusable apparels in their wardrobe. Instead of being disposed, these apparels can be donated to charities or it can be shared with their fellow family members. By understanding the huge environmental impact due to the textile disposal, many developed countries in the world are following the reuse method successfully in recent time. One of the successful method is, the Swap meets, it is a type of meet wherein participants exchange their valued but no longer used clothing for clothing they will use. Where people will exchange their clothing's, by doing so, the lifetime of the apparel extended and the environmental impact also. The U.S. Office of Resource Conservation and Recovery states that the average person throws away 60 pounds of clothing per year (McInerney 2009). While many of these textiles are recycled, almost 145 billion pounds are still routed to landfills every year. If 1000 people stopped dumping and started swapping, 11 tons of unnecessary landfill waste would be saved and treasured (McInerney 2009).

To encourage the people, these communities have also stated different websites for the clothing exchange process, so that people can trade their used apparel product along with some other apparel products of others. The websites like SwapStyle.com, Dig N Swap, ClothingSwap.com and etc. are working based on this reuse concept to reduce the impact of clothing disposal. The average consumption of fabric in the United Kingdom is 17.5 kg per person per year (McInerney 2009). The researchers also proposed that, out of 60% of the garments which ends up with landfill a significant amount of the apparels can be reused by initiatives such as clothes swapping, buying second hand and leasing expensive special occasion dresses. They have additionally noticed that individuals ought to be prepared to reuse their "never worn" thing as opposed to discarding (McInerney 2009).

Recycle—is the process to change items considered as waste into new products to prevent waste of potentially useful materials and reduce the consumption of fresh raw materials. Most of the time energy and water is used to change the physical properties of the waste material. Hence, it is recommended to go for reuse and reduce strategy. Recycle is concerned with providing the manufacturer with re-processed raw material to use as an input to make new goods. Recycling material saves resources and usually uses less energy than the production of new material. The recycling process can be performed into two ways namely as discussed in the following sections.

1. **Up-cycling**—It is the way toward changing over old or discarded materials into something valuable and often beautiful. While doing up-cycle, the process creates a new purpose to the old item. The objective of up-cycling is to avoid squandering possibly useful materials by making utilization of existing ones. This decreases the utilization of new raw materials while making new items. Lessening the utilization of new raw materials can bring about a decrease of energy usage,

contamination, ozone depleting substance discharges and so forth (Braungart and McDonough 2002). The process of up-cycling is intended to work contrary to shopper culture, urging individuals to consider new and creative approaches to utilize things, rather than just purchasing new merchandise. It likewise benefits the environment, by promoting reuse over disposing of at whatever point conceivable.

2. **Down-cycling**—A procedure of changing over important items into low-value raw materials. Contrasted with the up-cycling procedure, this one is least preferred. For instance: making recycled papers from paper, making rags from used cloths and apparels. Despite the fact that down-cycling helps the planet since it keeps things out of landfills (for a period at any rate) commonly it will in the long run wind up there.

6 Fabric Wastes in Various Part of Apparel Industry

In apparel industry, the initial process starts in the fabric department, where the manufacturing plant receives raw material from the supplier. This first point of waste generation is the fabric inspection. After inspection, the fabric is laid on the cutting table and cutting was carried out by cutting masters. It is the second point of waste generation in the department. After cutting process, the cut parts were numbered sequentially to avoid shade variation during the sewing process. After numbering and bundling process all the cut part components were transferred to sewing line for sewing process. This is the third point of waste generation. During sewing process, either due to sewing quality issues or cutting issues the components uses to get rejected. After sewing the garments were passes to end line quality auditing department, where the quality checkers will perform 100% end line checking, here if any un repairable defects found the garments will be rejected. This is another point of garment rejection. After this the garments will be packed and sent for dispatch. The following Fig. 5 represents the process sequence and Waste generation points in the apparel production process.

Figure 6 details the different types of wastes generated in the apparel industry from the sample making process to the retail shop. They are details as follows.

1. Sampling waste as Garment and Fabric yardage.
   In sampling department, during the garment development stages, different types of the fabrics will be sourced and utilized for the sample development. After development, the garments and fabrics will be sent to the buyer for the approval and based on the comments, the sampled or fabrics may be accepted and rejected. In spite of the acceptance, the buyer may also want to revise the style feature or incorporate few changes in the design. In that stage that samples and fabrics are thrown out as a waste at the least case.
2. Cutting department waste—The first waste developed by cutting department is cut waste, which was discussed in detail in the following section. The second

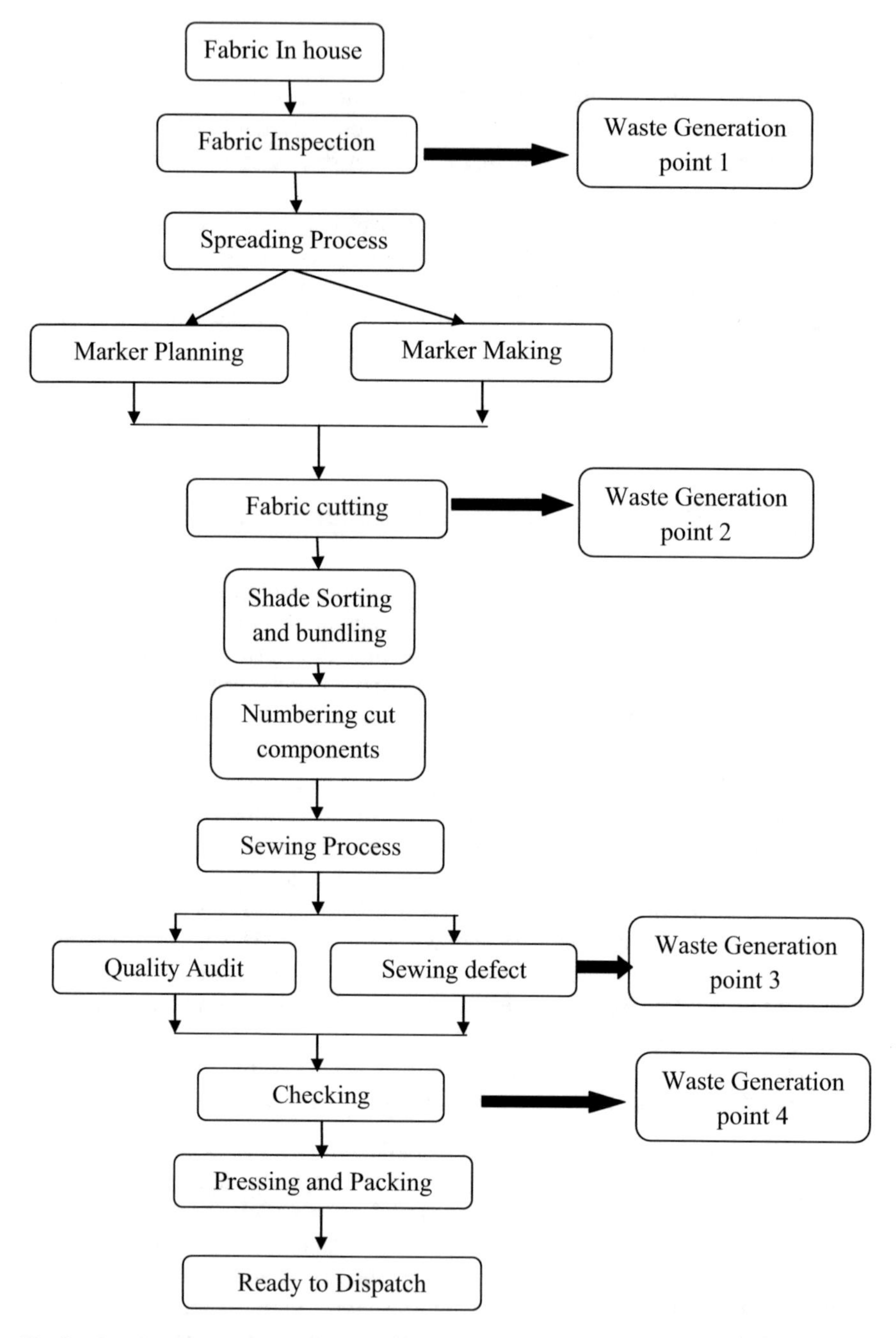

Fig. 5 Manufacturing sequence in apparel industry along with waste generation points different types of fabric from apparel industry

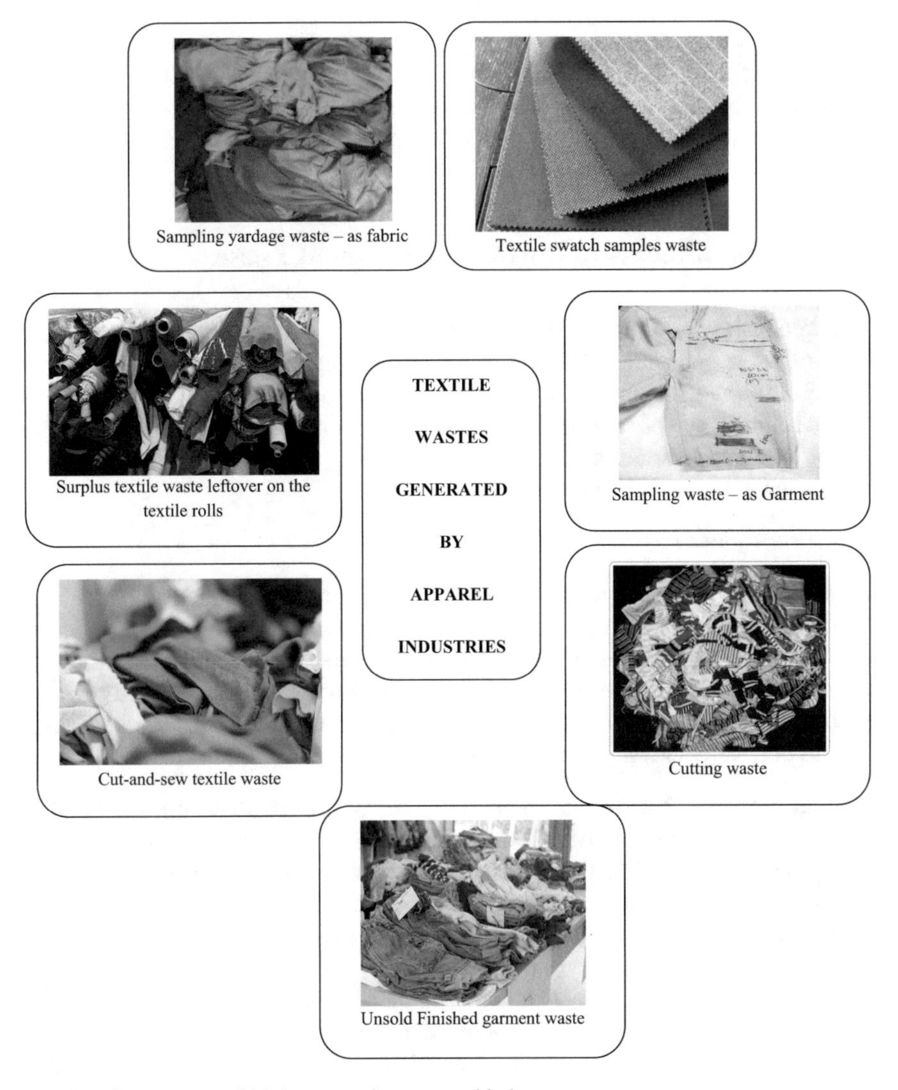

Fig. 6 Different types of fabric wastes from apparel industry

one is left over fabric after the required consumption. This leftover fabric to the maximum extent will be utilized and very short and unused material will be discarded as waste in this stage.

3. Cut and sew waste—This is the waste that generally coming out of sewing defect and quality issue. This waste are sometimes fully made garments or sometimes semi finished garment.
4. Unsold finished garment—It is another area, after the completion of the particular season, the manufacturer or the brand will come up with new set of styles

according to the current trend and the old style will be removed from the retail outlet. Most of the time these items will be sold as a second hand product or for the discounted prices. However, few brands because of their brand image they will shred or discard the old cloths as a waste.

7 Factors Influencing Cutting Department Losses

Cutting department of the apparel industry is one of the important departments in terms of waste generation. The aim behind the cutting department should be waste minimization and this can be achieved through many different ways. The most popular and adapted method in waste minimization is use of computer aided system. This section details the various cutting department activity which has major influence on the efficiency of the department.

1. **Marker planning**

Marker making is the way toward deciding the most efficient layout of pattern pieces for a specific style of fabric and size combination. Here, all the patterns of the selected styles and selected size combination are plotted in a paper, which is generally known as marker paper. This aids towards the better utilization of fabric and reduces the wastage to the maximum. There are two different methods used for the marker making process in apparel industry (Puranik and Jain 2017).

Manual method—This process required a skilled and experienced marker maker. The pattern arrangement and size combination ratios are decided by the marker maker and so the efficiency of the process majorly relies on the skill of the marker creator. This manual process can be performed either by drafting the original size patter on paper or by making a miniature size patter on paper and afterward duplicating the same on the fabric.
Computerized method—Application of automatic method using computer aided system increase precision, builds control over influencing factors and ultimately reduces the time significantly than the manual process. In this process the developed designs first entered into the memory of the computer aided system. Once the pattern details fed into the system, the marker plan will be generated either automatically or by automatic with user interface method. Out of these processes, the interative process is advantageous due to the use of operator skill and also the computer programming setup.

2. **Requirement of fabric in relation to garment style**

The waste generation in the cutting department is critically influenced by the design requirements of apparels. This is one of the major factor which influences the waste generation in the cutting department. For instance, if a specific apparel design required the fabric cut as cross grain, at that point the cloth wastage is unavoidable.

Same manner, the fabric will be wasted while a garment components should be coordinated with the style of the garment like, stripes, checked in pocket or collar getting matched with body style. Generally fabric width and length are the corresponding influencing factor for marker length and width. Markers are made according to the fabric width and the quantities of sizes. Hence, the percentage of waste increases, when the company has different fabric widths in a single order. In this case also, the waste can be reduced by grouping the similar width fabric together and also be generating a separate marker plan for different fabric widths (Solinger 1988). Another important aspect which majorly wastes fabric in a marker plan is, size and colour combination requirements.

3. **Marker efficiency**

Marker efficiency is another factor which has a major impact on the fabric waste developed in the cutting department. The efficiency of the marker can be calculated as follows (Solinger 1988).

$$\text{Marker efficiency} = (\text{Area of pattern pieces}/\text{Total fabric area}) * 100$$

Lowest fabric wastage can be achieved by increasing the marker efficiency. In general, the wastes in the marker planning are the parts which are not converted into garments. In other term, the fabric left in between the pattern pieces. The software from marker planning software's generally calculates the marker efficiency not only considering the wastes outside marker but also the tiny portions inside the marker too. The efficacy of the marker plan in generally influence by factors like fabric characteristics, shapes of pattern pieces, fabric utilisation standards and marker quality (Carr and Latham 1994).

4. **Other Fabric Losses in cutting process**

In both the above mentioned marker planning process, there are two different losses use to occur irrespective of the process. They are namely, Marking loss and Spreading loss.

1. Marking loss—It represents the waste fabric presents in the un usable area of the marker.
2. Spreading loss—It represents wastage of fabric outside of the marker plan.

The spreading loss can be broadly classified into different groups as in Fig. 7, namely ends of ply allowance, Splicing losses, edge losses, splicing losses, remnant losses, ticket length losses, etc., which are discussed below (Carr and Latham 1994; Kunz and Glock 2005).

i. **Ends of ply allowance loss**:
   For each spreading process, the end of the each side of the spread will be left without marking in view of the adaptability and extensibility characters of the fabric utilized. Usually, these allowances are set apart around 1 inch on each side, roughly, 2 inch in a single ply. In case of stable fabric like woven material, the

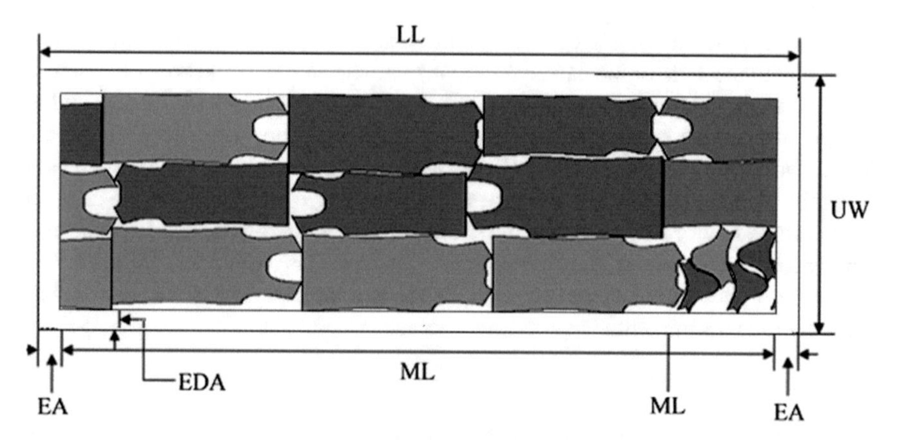

Fig. 7 Waste generated with the marker planning process. Where, EA = End allowances, i.e. allowances at the beginning and end of a layer; LL = Lay length, i.e. marker length + end allowances; EDA = Edge allowance, i.e. allowance at the fabric edge; ML = Marking loss, i.e. waste from within the lay plan; UW = Usable width, i.e. cloth width—edge allowance; and ML = Marker length

allowances can be reduced around 1 cm on each side based on the requirements. The research studies evaluated that the end ply losses alone wastes 1–2% of total order quantity approximately for a particular order. In general, the spread with higher length has less wastage rate than the spread with lower length (Kunz and Glock 2005).

ii. **Splicing Losses**:
Splicing is a process of joining two fabrics together. In cutting room, the marker planning is preferred only by considering the width of the fabric. The length of the spread will be adjusted based on the requirements in the efficiency. So the fabric length and the ply length cannot be matched exactly. Because of this reason, during spreading process, the fabric role may be emptied in the middle of the ply/spread. In that situation, both in manual and computer aided spreading the operator need to use/load up the new fabric roll. In that case the laying person will be making joint with the previous roll either by sewing or by overlapping the ply. In either case, the fabric shapes cut at those places are wasted. It is estimated that these loses varies from 0.5 to 1% of the total fabric usage.
The splicing process also performed during the situations like fabric defects. When ever, there is a objectionable fabric defect, like GSM cut hole, huge oil marks and etc. In this case these defects were cut and eliminated and after that the fabrics were spliced. The more frequent splicing increases the fabric wastages. Based on the quality policies and requirement of the customer the manufacturing firm may leave or cut out the defects on the fabric. This kind of losses can be reduced by legitimate planning and execution of quality policies in the firm.

iii. **Edge Allowances**:
In knitted apparel industry, the fabric width variation is a common issue. This is because the manufacturers use to develop fabric with different supplier or

different machines to achieve the target on time. In normal case, during planning process, to avoid small variation in the edge of the fabric the marker planning will be performed 1 cm on both side, meaning that 2 cm lesser width of the fabric. This width is called as usable width. Hence, when ever, there is a much variation noted in the fabric the loss will be considerably high. For example if a fabric with a width of 100 cm, used with 2 cm allowance totally, (1 cm both side) will be wasting 2% of the total fabric as an edge allowance. This wastage can be reduced by reducing the allowance from less than 1 cm, if the fabrics are with the same width and more stable. Hence, the direct loss by edge allowance can be reduced by proper planning in the fabric manufacturing. Thus the width variation between the fabric rolls can be reduced an so the waste.

iv. **Leftover Losses**:
Remnant lengths or left over lengths are another types of waste which mostly generated in the cutting department of the apparel manufacturing unit. When ever, the company orders fabric with excess quantity and company separates the fabrics based on fabric shades, the remnant length will be generated commonly. These fabrics are utilized for garment cutting if it had enough length or enough number of roles, then those fabrics will be utilized as a separate lay or as a stepped lay along with the main lay. However, the utilization percentages will less due to the wide variation in the remnant fabric.

v. **Ticket Marking Losses**:
The woven fabric and knitted fabrics, during sales phase, the fabrics were measured for their length in case of woven and weight in the case of knitted fabric and any other technical details required for the identification of the fabric will printed in the fabric directly or ticket format paper will be stuck on it or Permanent fabric markers will be used to write these info's on the fabric. In many cases, fiber type, GSM, Total length will be mentioned in the content. If these kinds of process are properly monitored then the losses can be reduced (Carr and Latham 1994; Kunz and Glock 2005).
The fundamental aim of this study is to identify the different fabric wastes developed from the apparel industry and to identify the major point of waste generation in apparel industry. Further the research is also focused towards the collection and utilization of wastes from apparel industry. This recycling process develops a new fabric from the waste fabric collected from industries. In this point of view a case study was conducted various apparel manufacturing industries in Tirupur.

8 A Case Study in Apparel Industry

As a preliminary work, the previous researches on the apparel industry wastes were analyzed. There are very few studies are available in this area. A study by Rahman and Haque (2016), showed the different waste percentage of the cutting department of an apparel industry. They have studied the waste fabric production in the cutting,

Table 1 Percentage waste in different sections of apparel industry (Rahman and Haque 2016)

Day	Fabric weight taken in grams	Fabric waste %				Total waste %
		Cutting	Panel checking	Sewing	Finishing	
1	16,551	16.5	6.67	4.15	1.66	28.98
2	25,255	15.04	6.78	4.24	1.69	27.75
3	27,541	13.83	6.90	4.29	1.72	26.74
4	27,798	10.96	7.12	4.45	1.78	24.31
5	29,717	11.56	7.09	4.42	1.76	24.83
Average	**25,372.4**	**13.57**	**6.91**	**4.31**	**1.72**	**26.52**

panel checking process, sewing and finishing process. The researchers conducted a study on five different knit T shirt manufacturing apparel unit in Bangladesh and the average waste percentage calculated department wise in those industry. In their study they have mentioned that out of the selected fabric quantity, in and average a 26.52% of fabric goes as a waste in the overall garment manufacturing process. They have also mentioned that out of that 26.52%, a maximum of 13.57% of waste generated in the cutting section. The average wastes generated at the end of each department activity are provided in the Table 1 (Rahman and Haque 2016).

In an another study performed by Tanvir and Mahmood (2014), they have analysed the knit garment manufacturing industries in the Bangladesh. They have measured the waste generation quantity in each point of the apparel industry for different order. They have collected the data from 25 industries all over the Bangladesh. As a finding they have listed the following Table 2 in their result (Tanvir and Mahmood 2014).

From the study conducted in the 25 apparel industry, researchers collected the waste percentage in the four points namely, inspection loss, cutting loss, sewing loss and finishing loss. From the study they had concluded that around 25% of the fabrics loss happening in the process. Their results also indicated that, out of all the selected industries, the waste percentage was noted higher in fabric inspection and fabric cutting activity. The waste generation the cutting section is significantly higher than that of other departments of the apparel unit. In this study it can be seen that out of the selected fabric quantity of 136,930 kg, the maximum of 2910 kg of fabric wasted in the cutting section alone (Tanvir and Mahmood 2014). Based on the available literature, it can be clearly seen that, the major part of the waste from the industry is from cutting department. While considering the waste from the sewing and finishing department, the cutting department wastes are significantly high. In view of these research, the cutting waste percentage of the Tirupur knit apparel industries were collected to understand the approximate contribution percentage. The results of the survey are provided in the Table 3.

Table 2 Total waste percentage calculation in different industry (Tanvir and Mahmood 2014)

Factory number	Input fabric in kg A1	Waste generation in Kg					Waste percentage (%)
		Inspection loss	Cutting loss	Sewing Loss	Finishing loss	Total Waste	
		Point 1	Point 2	Point 3	Point 4	A2	
1	700	35	50	20	10	115	16.25
2	750	30	40	25	15	110	14.67
3	780	40	50	15	10	125	16.03
4	800	25	30	30	20	105	13.13
5	820	20	45	30	15	110	13.42
6	880	25	40	35	20	120	13.63
7	910	50	70	30	25	175	19.24
8	950	45	65	25	20	155	16.34
9	990	25	35	35	15	110	11.12
10	1000	50	50	30	10	140	14
11	1100	25	40	25	5	95	8.64
12	1900	100	100	50	40	290	15.27
13	2000	80	60	30	50	120	6
14	2300	110	100	50	20	280	12.18
15	2500	25	20	10	5	60	2.4
16	3000	20	40	30	10	100	3.34
17	3200	60	35	20	20	135	4.26
18	3600	50	30	10	15	105	2.9
19	3900	90	35	30	20	175	4.49
20	4000	80	30	25	25	160	4
21	4100	40	25	50	20	135	3.30
22	4250	35	30	30	10	105	2.48
23	4400	55	25	50	5	135	3.06
24	4700	70	30	30	5	135	2.89
25	5000	65	25	50	10	150	3
26	14,000	50	120	20	45	235	1.68
27	1100	25	15	25	10	75	6.8
28	24,200	220	200	50	40	470	2
29	23,100	140	180	45	30	385	1.6
30	1600	10	10	25	5	50	3.1
Total	**136,930**	**1585**	**1325**	**930**	**540**	**4240**	

Table 3 Cutting department waste calculation

Order No	Gram per square meter of the fabric	Fabric input quantity in Kg	Fabric waste quantity in Kg	Waste percentage
1	155	850	59.5	7.00
2	155	1206	237	19.66
3	175	1885	482.42	20.38
4	155	2000	415	20.77
5	184	1278.59	464.77	26.66
6	175	1666.59	401.19	19.4
7	208	2207.64	591.90	21.14
8	160	1260	340	26.9
9	180	1010	240	23.7
10	200	1450	648	44.6
11	155	1850	385	20.8
12	170	2250	462	20.5
13	180	1885	420	22.2
14	180	2020	382	18.9
15	175	1550	216.5	13.9
Average		**2436.82**	**5745.28**	–
Average waste percentage		**23.57%**		

From the result it can be seen that for 15 different orders from 10 various factories, the average waste percentage is noted as 23.57% in cutting department alone. These results are in line with the studies of the pervious researchers Tanvir & Mahmood (2014) and Rahman & Haque (2016). Another important factor observed during the study is, the factories where we collected the cutting waste uses computerized marker plan. It can be seen that, the influence of garment style and style requirement plays a vital role in the fabric waste irrespective of the methods used for marker planning. Particularly in some style, example order number 10, in the Table 3, represents a style which wastes around 44.6% of the fabric only due to style requirements.

9 Materials and Methods

9.1 Material

The raw material used for the fabric production is fabric wastes from the cutting department. The collected fabric wastes were reconverted into fibers using willowing machine. The waste collected from the industry was manually segregated for colour similarity. About 50 kg of the collected fabric were fed into the willowing

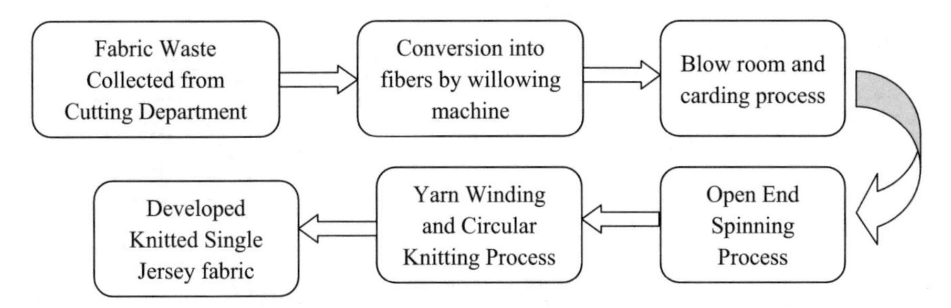

Fig. 8 Process flow of the fabric development process

machine. The willowed material was the passes to different yarn formation process as mentioned in Fig. 8.

9.2 *Methods*

9.2.1 After treatment for the developed fabric

The fabric developed from the cutting waste recycling has very rough surface, due to the different fiber lengths, count mix up and other processing factors like, carding and open end spinning process. Hence to improve the handle of the fabric, the fabric is treated with enzyme and also finished with cationic silicone softener.

Enzymatic treatment—For the finishing process the enzymatic desizing agent is added into the bath in the proportion of 0.6 g/L along with the detergent proportion of 0.5 g/L. The process is performed 20 min, 50 °C temperature with the material to liquor ratio of 1:10. After enzymatic treatment the fabric is neutralized with rise process.

Softening process—For the Softening process the cationic softening agent is added into the bath in the proportion of 1 g/L along with the silicone proportion of 0.5 g/L with the material is to liquor ratio of 1:8. The process is performed 20 min with 40 °C temperature. After finishing, the fabric is dried and calendared before taken to the testing purpose.

9.2.2 Performance analysis

The developed fabrics were analysed for the following tests and compared with the normal 100% cotton single jersey fabric. The testing methods and the standards used are detailed as follows (Savelle 1999).

- Dimension Stability—AATCC 135-2015
- Bursting Strength—ASTM D3786

Table 4 Properties of recycled and normal yarn

S. No	Description	Normal cotton yarn	Recycled cotton yarn
1	TPI	22.1	18.21
2	Thin places/Km (−50%)	0	27
3	Thick places/Km (+50%)	8	42
4	Neps (+200%)/Km	32	80
5	Hairiness index	4.1	4.8
6	Strength cN/tx	17.5	12.1
7	Irregularity CV%	17.88	28

- Pilling resistance—IS:10971
- Abrasion resistance—ASTM D4966
- Spirality—AATCC 179-2004

9.2.3 Product development and analysis

The developed fabric was utilized for the garment development process. The designs were created for the garments and developed garments also anlaysed for their production cost.

10 Results and Discussion

As mentioned earlier, the collected cutting waste from the apparel manufacturing unit was shredded into fibers and spun back into yarn and finally the knitted fabrics were developed using weft knitting machine. The developed fabrics were analysed for their suitability as an apparel fabric.

11 Yarn Properties

The yarn properties of normal cotton and recycled cotton yarns are provided in Table 4. From the result it can be seen that the imperfection are noted higher in the case of recycled yarn than the original yarn. The strength of the recycled yarn is comparatively low than the normal yarn.

Table 5 Fabric properties of the normal and recycled fabric

S. No	Description	Normal fabric	Developed fabric
1	Yarn count	30s Ne	28s Ne
2	Yarn type	100% cotton	OE Yarn 100% Cotton dyed yarn
3	Loop length	27 mm	32 mm
4	Diameter	30 mm	26 mm
5	GSM	160	177
6	Thickness	0.52 mm	0.64 mm

Table 6 Dimensional stability of the developed fabric in width and lengthwise direction

S. No	Lengthwise shrinkage %		Widthwise shrinkage %	
	Normal fabric	Developed fabric	Normal fabric	Developed fabric
1	2	4.2	0.5	0.9
2	2.2	4	0.9	0.8
3	3.5	4.2	0.6	1
4	2	4.5	0.5	0.9
5	2	4.2	0.6	1
Mean	**2.34**	**4.22**	**0.62**	**0.92**

12 Fabric Properties

The properties of the developed fabric from the recycled yarn were compared with the normal cotton yarn made fabric and the details of the developed fabric were listed in Table 5.

13 Physical Property Analysis

13.1 Dimensional Stability

Dimensional stability refers to a fabric's ability to resist a change in its dimensions. The processes like washing, drying, steaming and pressing will alter the dimension of the garment either by shrink or by growth base on the type of fabric and the results of the developed fabric was provided in Table 6.

The test results of the developed fabric shows higher shrinkage percentage than the normal fabric in the both the direction. In the case of the lengthwise shrinkage, the developed fabric had a percentage of 4.22 as an average but the normal fabric had only 2.34%. The developed fabric had approximately double time shrinkage than the normal fabric. The reason for higher dimensional instability in the developed fabric may be due to the usage of fiber mixture in the open end spinning process. It is observed that generally the structural differences and fiber type play a large part in determining the dimensions of these fabrics. In particular, with respect to the knits structure the blend yarns have a lower dimensional stability compared to fabrics from 100% cotton ring and open-end spun yarns (Nihat Çel and Çoruh 2008). In this case, the selected recycled yarn is very similar to the blended yarn, where it was blended with different types of cotton fiber. Hence, the results are expected as lower dimensional stability than the normal cotton fabric. These findings are in line with the findings of Onal and Candan (2003).

In specific to the widthwise shrinkage, the results of the current study was in disagreement with the previous research works. The researchers mentioned that the open-end rotor fabrics shrink somewhat less in widthwise than ring spun samples for plain knits, because they are more dimensionally stable due to better extensibility (Erdumlu1 and Ozipek 2009). The contradiction in the findings of the current research may be due to the blending of different quality of fiber and length variation in the cotton fibers from recycling process. However, these finding falls in line with the findings of some other researchers, who mentioned that the higher shrinkage tendency of the open end rotor knitted fabrics (Burnip and Saha 1973; Lord et al. 1974; Sharma et al. 1986). As an overall, the dimensional stability of the recycled cotton yarn is observed less than the normal 100% pure cotton yarn.

13.2 Bursting Strength

The strength of the fabric is an important parameter in all aspects. The main requirement is the fabric must have the required strength to pass through all the manufacturing process (Ertugrul and Ucar 2000). When a fabric fails during a bursting strength test it does so across the direction which has the lowest breaking extension. This is because when the fabric tested like this, the same amount of stress applied all the direction of the fabric.

In this research, the bursting strength results of the knitted fabric evaluated for normal cotton yarn and the recycled, open end spun knitted fabric. The results revealed that bursting strength of the developed recycled cotton yarn fabric 2.5 times lower than the normal yarn. Whereas the normal 100% cotton fabric has a strength of 160.8 PSI as an average but in the case of the developed fabric it is noted as 63.6 PSI as an average of five reading. There are two possible reasons for this strength reduction in the knitted fabric. The first one is, the yarn formation system used. Researchers mentioned that the bursting strength of the fabrics produced from ring spun yarns were generally higher compared to the other systems like Open end and Vortex spinning. And this is expected due to the lower tenacity values of the open end rotor spun yarns compared to the ring spun yarn (Erdumlu1 et al. 2009).

The second possible reason is the fiber content. As far as fiber content is concerned, the highest bursting strength values were obtained in 100% cotton fabrics. The results

Table 7 Bursting strength value of recycled fabric

S. No	Normal fabric (PSI)	Developed fabric (PSI)
1	160	63
2	162	66
3	161	65
4	160	63
5	161	61
Mean	160.8	63.6

Table 8 Pilling resistance values of recycled fabric

S. No	Normal fabric	Developed fabric
1	3	1
2	4	2
3	3	1
4	3	1
5	3	1
Mean	3.2	1.2

of Candan et al. (2000) was in line with the findings of the current research. Even though the developed fabric has 100% cotton yarn, the properties of the individual recycled fibers are comparatively inferior to that of 100% virgin cotton. This could be the possible reason for the reduction in the bursting strength value of the developed fabric. The yarn counts and their strengths are also important parameters determining the fabric strength, from the yarn analysis, it can be seen that recycled yarn has lower strength compared to the normal cotton yarn. The bursting strength results of developed fabric is provided in Table 7.

13.3 Pilling Resistance

Pilling is a phenomenon of fiber movement or slipping out of yarns, which is usually happening on the fabric surface during abrasion and wear. The results of the current research were given in the following Table 8.

The results indicate that the developed fabric showed poor performance against the pilling resistant. It is noted that the pilling resistance of the normal ring spun cotton yarn knitted fabric had a rating of 3 as an average which represents the moderate pilling behaviour. And the developed fabric from the recycled yarn showed very low resistance to the pilling. The rating is provided as 1, meaning that the developed fabric has showed severe pilling formation on the surface. In general, as discussed by Candan et al. (2000), the knitted fabric made off open end yarns perform better than the knit fabrics made from the ring spun yarn. The possible explanation for such pilling behavior is that the well aligned, compact structure of ring spun yarns, compared to the loosely wrapped outer layer of open-end yarns with trailing loops, does not allow easy fiber pull-out and fuzz removal, which contributes to the lower pill ratings. However our current research findings are in contradiction with that where the open end yarn formed fabric performance noted poor. The main reasons behind the results are recycled fiber. Since the recycled fibers are not uniform in length, the number of loose fibers at the end increases. This might be the main factor for the high dense pill. Along with the recycled un-uniform fiber length, the lower twist level less inter fiber friction. Shorter fiber altogether caused higher tendency to pill with the developed fabric.

13.4 Abrasion Resistance

Abrasion resistance of the fabric and pilling performance of a fabric are interrelated. A higher number for the abrasion resistance means a higher abrasion resistance, but a lower number for pilling means a higher pilling performance. Because of this, it can be observed from the results that the abrasion resistance and pilling performance of the recycled open end yarn fabrics has a maximum value (Can 2008). The results from the current research revealed that the abrasion resistance of the open end spun recycled cotton yarn has abraded higher than that of the ring spun normal yarn. This is due to the nature of fiber and yarn characteristic. The long fibers inside a yarn or fabric confers an improved abrasion that the fabric or yarn with short (Savelle 1999).

In this case the recycled fiber, lengths are highly varied than the normal cotton and also the twist factor. Compared to the ring spun normal yarn the open end recycled yarns have lower twist factor. The better abrasion resistance can be obtained with the optimum amount of twist in a yarn. When the twist level is less, the fibers from yarn structure easily raveled out and reduce the abrasion resistance. These are the main reason for the slightly reduced abrasion behavior of the recycled cotton fabric. The average weight loss percentage of the normal fabric found to be 8.1% and in the case of the developed fabric it is noted as 9.2% (Table 9). These results were also in line with the findings of the previous workers' results (Lord et al. 1974; Alston 1992).

Table 9 Abrasion resistance (weight loss percentage) value of recycled fabric

S. No	Normal fabric (%)	Developed fabric (%)
1	8.3	8.5
2	8.1	9.2
3	8.7	9
4	8.3	8
5	8.1	9.2
Mean	8.3	8.7

13.5 Spirality

It is one of the dimensional mutilation issues. In plain knitted fabric produced from circular knitting process usually suffer by this issue. The spirality problem affects both the appearance and performance of the knit apparels. The garment with this issue usually identified by the displacement of side seam to the front and back side, this is one of the major quality issues in the apparel manufacturing process. This occurs due to the displacement of wales and courses in the knitted fabric in an angular manner from its original perpendicular angle (Afroz et al. 2012; Hossain et al. 2012).

The analysis of spirality angle of the knitted fabric is important in the quality point of view. The results of the current research revealed that the spirality value of the developed recycled knitted fabric observed comparatively high than the normal 100% cotton knitted fabric. This is expected results based on the previous observations, the open end spinning process and the recycled fiber characteristics are the main reason for the results. This was totally in contradiction to the findings of the previous researchers, where they have mentioned that the twist direction and amount of the twist are the major reason for spirality. In open end spinning process the twist level is lesser than the ring and compact spinning and so the torque on the yarn also less during fabric formation. Thus the open end yarn has a reduced spirality (Mezarciöz and Oğulata 2011). Hence, it is clear from the results that the higher spirality in the developed knitted fabric is totally associated with the characteristic nature of the recycled fiber. This could be the only reason that the open end spun yarn has the higher spirality than the ring spun yarn (Hassan 2013). The spirality values of the developed fabric are provided in Table 10.

13.6 Cost Analysis of the Developed Fabric

The cost analysis results of the developed fabric and the normal cotton material are provide in the Table 11. It can be seen that the total cost of the 1 kg normal fabric is Rs. 374 and in the case of developed fabric it is Rs. 254. The reduction in cost is mainly associated with the yarn and dyeing cost.

Table 10 Spirality (%) value of recycled fabric

S. No	Normal fabric (%)	Developed fabric (%)
1	7.1	10.3
2	7.5	10
3	7.6	9.7
4	8.3	10.2
5	8.5	9.6
Mean	7.8	9.96

Table 11 Cost analysis of normal and recycled fabric

S. No	Cost factor	Normal fabric cost in Rs. per Kg	Developed fabric cost in Rs. per Kg
1	Yarn cost	230	170
2	Knitting cost	17	17
3	Dyeing cost	120	–
4	Washing	–	60
5	Compacting and finishing	7	7
	Total	374	254

14 Product Development

With the help of the developed fabric, the men's and women's wear apparels were developed as shown in the Figs. 9 and 10.

15 Sustainable Brands on Focus

Many leading brands are now considering the impact of their production process on the environment. Hence, they have initiated recycling or reusing process into their manufacturing facility to reduce their environmental impact. The initiatives of the few brands, on sustainability are provided with their adapted method.

- **Evrnu™** has made an imaginative new innovation that reuses cotton clothing waste to make premium, sustainable fiber called as "pristine new fiber". The company basically converts all collected solid wastes into fluid and then processes into unadulterated fiber that can go up against the requirements of the designers and product development team. They had likewise asserting that they are making this with 98% less water than it takes to make conventional cotton fiber and with 90% lessened CO_2 emanations contrasted with polyester generation. The production process of Evrnu brand is provided in Fig. 11.
- **Fabscrap** offers easy pickup and recycling of textiles in New York City. They had tie up lots of leading fashion brands, independent designers, cutting rooms, textile arts organizations, schools, and regional processors. They intend to develop a system of up-cycling and down-cycling organizations to guarantee greatest diversion from landfill.
- **ReRoll**—ReRoll creates flat textile goods made from cutting room scraps. Scrap materials are gathered from manufacturing plants, production facilities and designers and the procedure means to end the inefficient practices of the garment industry. It pays attention towards transforming fabric "scraps" or "waste" into resource or a raw material for new production. The studio was established by creator Daniel Sil-

Fig. 9 Men's round neck t Shirt with chest print

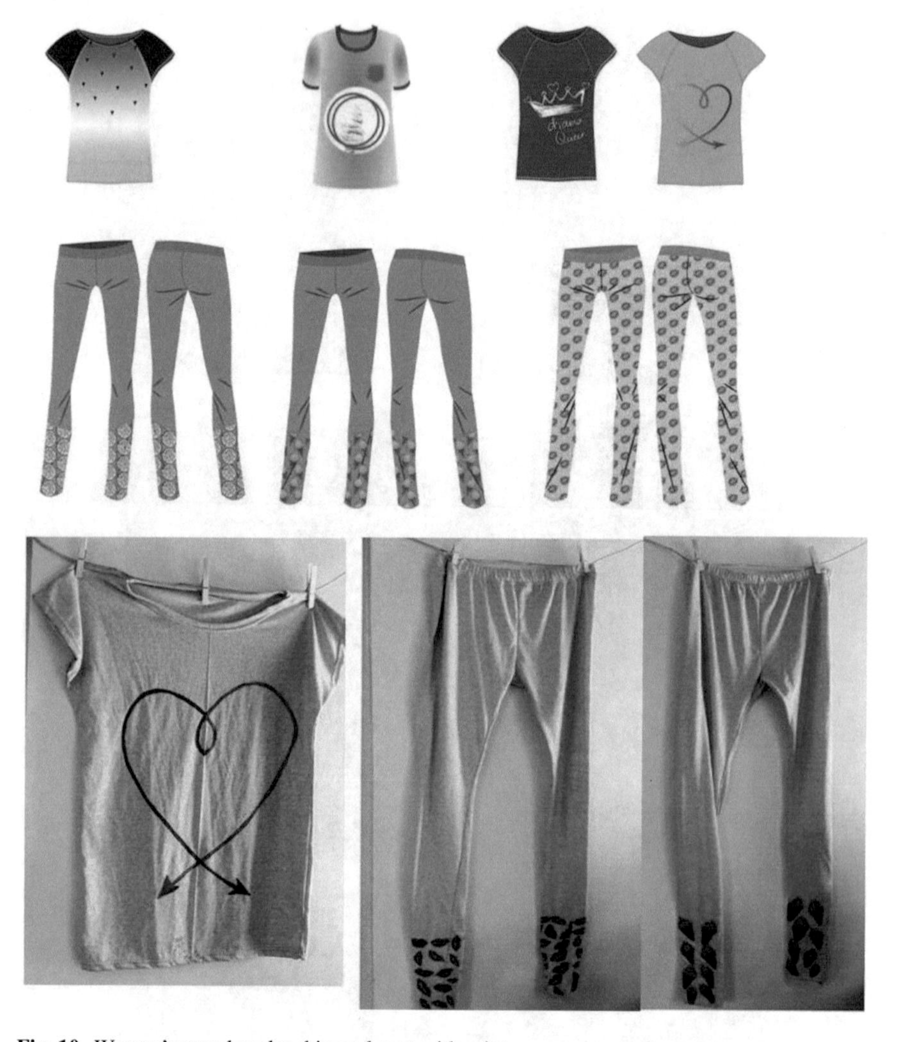

Fig. 10 Women's round neck t shirt and pant with print

verstein. His point, the Zero Waste is a mission to kill inefficient form of industry standards by alternative fashion made using his unique procedure.

- **TONLE'**—The organization begins their procedure with scrap sourced from mass textile manufacturing industries. They make handcrafted apparel and adornments marked by their Cambodian producers. But still, Tonlé working hard to take out their last 2–3% of waste in processing, so that they can make the organization really zero waste industry. The greater part of the process is they convert their pieces of fabric waste which cannot be used in design, as recycled paper. These papers again used to make the "Tonlé" hangtags.

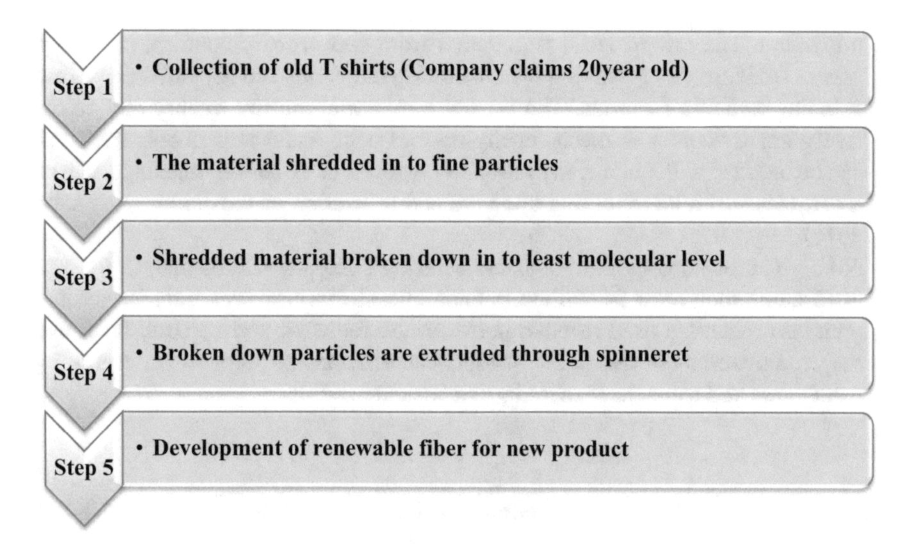

Fig. 11 Process cycle of Evrnu brand recycling activity (*Source* www.evrnu.com)

- **Ecotec**—Is an Italian company, which develops recycled yarn from pre dyed cotton textiles, sourced as a scrap out of fashion and clothing manufacturing house. They also produce traceable yarns. Traceable yarns means, the customer learn about the farm/industry and the source of the farm/industry that produced the yarn, by simply entering yarn information printed on the lable. The company claims that they consume 77.9% less water than the similar textile manufacturing companies. This is due to the coloured yarn production, which totally eliminates the dying process, there by the manufacturer reduces the environmental impact.
- **Trmtab**—is one of the apparel trim recycling company which works on leather scraps from production facilities in India, specifically the leather trims that are the byproduct of leather goods production, and turns them into woven and chevron stitched accessories like wallets and clutches. The manufacturing industries reject the leathers with scares, marks, and other natural defects. These kind of good leathers were graded as second and third class only due to their look and appearance. This kind of leather waste will be recycled into useful fashion accessories by the Trmtab.
- **Patagonia** has for quite some time been a pioneer in sustainable textiles. They manufactured first outdoor clothing by consolidate reused plastic containers into their line of fleece products, in 1993. Now the company uses their processing plants to repurpose cotton scraps, converting into fabric and apparels. The organization guarantees the scraps from 16 virgin cotton shirts can be transformed into one recovered cotton shirt. Patagonia likewise utilizes recovered wool and reused nylon down.
- **Pure waste textile**—They assert that for creating 1 kg of cotton it required 11000 L of water and by doing reuse they diminish the effect of cotton waste and water

utilization. The raw material is gathered from two primary sources, (1) cutting wastes from article of clothing manufacturing industries and (2) yarn scraps from spinning/weaving factories. The wastes were sorted out by quality and colour. Every quality/colour is mechanically opened once again into fibers. The fibers can be blended with chemically recycled polyester or viscose filaments to achieve a particular usefulness relying upon the end utilization of the clothing (Chhabra 2016).

- **Nike**—The leading sportswear manufacturing company claims that, in the year 2015 alone they have 54 million pounds of factory scrap was transformed into premium materials used in Nike performance footwear and apparel. Nike uses recycled materials in their 71% of footwear and apparel products, in everything from yarns and trims, to soccer kits and basketball shoes.

16 Future Scopes for the Sustainable Production

The real issue with the present design drift is low quality fabrics. Implying that the pieces of clothing going to wind up as a landfill or as a down-cycled item at the most punctual. Despite the fact that it is smarter to down cycle the item, rather than sending them specifically to the landfill, the down cycled squanders or material will discover their approach to landfill soon. Everything is separated further and further until the point when it in the end achieves the landfill. The main conceivable innovation that will help the clothing industries is closed loop technology. Implying that, an item is reused once more into practically a similar item with same value. It is the fundamental point of the sustainable production since it basically copies the common procedure of life. The plant grows out of dirt, dies and incorporated back into the dust and the new plant grows from the same dirt, and the process generates no waste. It is the main plausibility for the mold business to create zero waste items and make nothing goes to the landfill, similar to the polyester fiber. Once the polyester filaments manufactured, they woven into a material, made into a piece of clothing, at the end broken into unadulterated polyester and woven into a material once more with same value, this is the form, which the fashion industry for the natural fiber.

However, from the current research, it can be noted that the recycling of the cotton fiber, could help to make closed loop technology in fashion industry. From the property analysis of the developed fabric it can be clearly understood that the recycling process, mechanically degrades the fiber property. These findings were in line with the report by the Sustainable Apparel Coalition, who mentioned that the closed loop technology is merely not possible in the natural fiber. They have mentioned that once cotton is dyed, treated or blended with other materials, the process no longer works. But, a hopeful note appeared in May 2016, when Levi's debuted a prototype of jeans in partnership with the textile technology startup Evrnu, made with a mix of virgin and chemically recycled cotton from old T-shirts. They had mentioned that the developed fabrics were not sensitive to some of the dyes, however, the developed a 100% cotton fabric from post consumer waste material

(Wicker 2016). In spite of the natural fiber, the application of closed loop technology is far away for the synthetic fiber also. Pantagonia successfully converted polyester into its core component and again into a thread. However, the brand is doing it for principle they had taken and not for the profit. They had also mentioned that the required high quality raw material as an input. This is another issue with respect to the synthetic material along with cost factor.

While talking about the sustainable manufacturing it is also important to know the customer acceptability of the recycled product. Recent studies have reported that consumers are willing to purchase recycled textile products and textile manufacturers are willing to use recycled fibres (Grasso 1996). Out of that, the findings supports that most of the customers shows like tendency towards recycled material are young and at college going age (Mahajan and Grasso 1991). Grasso recently reported that US textile manufacturers are recycling or reclaiming internally re-workable fibre, yarn, and/or fabric. In addition, there is a willingness on the part of textile manufacturers to produce products using recycled textile materials if it is economically feasible from recycled fiber. A report released by McKinsey & company listed few steps for the textile manufactures and consumers. They have mentioned the following points as preventive measures to make fast fashion more sustainable (Cheeseman 2016):

- Develop guidelines and practices for designing the clothing material that can be effectively reused or recycled.
- Invest in growing new fibers that will bring down the ecological impact of making textiles.
- Encourage customers to look after their garments in ways that will prolong in their utilization, for example, washing them in icy water.

17 Conclusions

The growth of readymade garment sector and increased per capita consumption of the garment along with the Fast fashion trend plays a vital role in the sustainability issues in the apparel sector. Few developed nations and manufacturing brands has taken initiatives towards the sustainable production. However, it can be seen that all the initiatives are in vein unless the customer understands the necessity of the sustainable product. The reuse and recycling are found to be the better option to reduce the environmental impact. But, the acceptance of the recycled product or an old material is still in lagging among the customer. As mentioned earlier, closed loop technology in the apparel industry is merely a decade away to be in practice, due to the technology availability and cost factor. The second important issue with the recycling products is the poor quality. Current research findings also outlined more that quality issue in the recycled fabric. In this work, we had developed garments with lower price compared to the normal garment, however, the quality aspect like, physical and comfort properties, the fabric is far low than a normal ring spun virgin cotton. The performance of the fabric in terms of bursting, abrasion, pilling, dimensional stability and spirality were analysed and noted lower than ring spun yarn. The major aim of the work accomplished by converting the wastes into recycled product, there by reduced the waste management expense. Even at low cost the product will give some revenue to the industry, in spite of spending money for waste disposal. But, the acceptance of the product in the market is in the hands of customer. Hence, the consumer mind set toward the sustainability is much more important in the current scenario.

References

Afroz N, Alam AKMM, Mehedi H (2012) Analysis on the important factors influencing spirality of weft knitted fabrics. Inst Eng Technol (IET) 2(2):8–14

Alston PV (1992) Effect of spinning system on pill resistance of polyester/cotton knit fabrics. Text Res J 62(2):105–108

Annual Report 2016–17 (2017) Ministry of textile, Government of India. http://texmin.nic.in/sites/default/files/ar_16_17_ENG.pdf. Accessed on Sept 2017

Apparel export promotion council-Annual report 2016–17 (2017) page 5–7. http://aepcindia.com/news/all-india-rmg-exports. Accessed on Sept 2017

Braungart M, McDonough W (2002) Cradle to cradle: remaking the way we make things. North Point Press, NY

Burnip MS, Saha MN (1973) The dimensional properties of knitted cotton fabrics made from open-end spun yarn. J Text Inst 64:153–169

Can Y (2008) Pilling performance and abrasion characteristics of plain-weave fabrics made from open—end and ring spun yarns. Fibres & Textiles in Eastern Europe 16(1(66)):81–84

Candan C, Nergis UB, Iridag Y (2000) Performance of open-end and ring spun yarns in weft knitted fabrics. Text Res J 70(2):177–181

Carr H, Latham B (1994) The technology of clothing manufacture, 2nd edn. Blackwell Scientifi c, Oxford

Cheeseman G-M (2016) The high environmental cost of fast fashion. http://www.triplepundit.com/2016/12/high-environmental-cost-fast-fashion/. Accessed on Sept 2017

Chhabra E (2016) India: two entrepreneurs turn waste into a business. http://pulitzercenter.org/reporting/india-two-entrepreneurs-turn-waste-business. Accessed on Sept 2017

Dwyer J (2016) A clothing clearance where more than just the prices have been slashed. http://www.nytimes.com/2010/01/06/nyregion/06about.html?mcubz=0. Accessed on Sept 2017

Erdumlu1 N, Ozipek B (2009) Investigation of vortex spun yarn properties in comparison with conventional ring and open-end rotor spun yarns. Text Res J 79(7):585–595

Erdumlu1 N, Ozipek B, Oztuna S, Cetinkaya S (2009) Investigation of vortex spun yarn properties in comparison with conventional ring and open-end rotor spun yarns. Text Res J 79(7):585–595

Ertugrul S, Ucar N (2000) Predicting bursting strength of cotton plain knitted fabrics using intelligent techniques. Text Res J 70(10):1–4

Fabric usage and various fabric losses in cutting room, processing, dyeing & finishing, features. http://www.indiantextilejournal.com/articles/FAdetails.asp?id=1307. Accessed on Sept 2017

Fast fashion quick to cause environmental havoc. http://www.uq.edu.au/sustainability/fast-fashion-quick-to-cause-environmental-havoc-143174 (2016)

Fletcher K (2008) Sustainable fashion and textiles: design journeys. Earthscan, Oxford

Grasso MM (1996) Recycling fabric waste—the challenge industry. J Text Inst 87(1):21–30

Hassan NAE (2013) An investigation about spirality angle of cotton single jersey knitted fabrics made from conventional ring and compact spun yarn. J Am Sci 9(11):402–416

Hossain M Md., Jalil MA, Saha J, Mia M Md., Rahman M Md. (2012) Impact of various yarn of different fiber composition on the dimensional properties of different structure of weft knitted fabric. Int J Text Fashion Technol (IJTFT) 2(1):34–44
http://www.wisegeek.com/what-is-upcycling.htm. Accessed on Sept 2017
https://www.evrnu.com/technology/. Accessed on Sept 2017
http://fabscrap.org/charlotte-index. Accessed on Sept 2017
http://zerowastedaniel.com/. Accessed on Sept 2017
https://tonle.com/pages/zero-waste. Accessed on Sept 2017
http://www.trmtab.com/about/#about-upcycling. Accessed on Sept 2017
http://bkaccelerator.com/interview-positive-impact-awards-jill-dumain-director-of-sustainable-strategy-patagonia/. Accessed on Sept 2017
http://purewastetextiles.com/. Accessed on Sept 2017
https://about.nike.com/pages/environmental-impact. Accessed on Sept 2017
http://www.ciromero.de/fileadmin/media/informierenthemen/gruene_mode/Jungmichel._Systain.pdf. Accessed on Sept 2017
http://www.ecotecproject.com/english.html. Accessed on Sept 2017
https://wayfaringyarns.com/2015/11/01/trace-your-yarn/. Accessed on Sept 2017
Jake Hall (2016) What is fast-fashion actually doing about sustainability? http://www.refinery29.com/2017/06/159074/fast-fashion-hm-transparent-sustainability. Accessed on Sept 2017
Joy A, Jr Sherry JF, Venkatesh A, Wang J, Chan R (2012) Fast fashion, sustainability, and the ethical appeal of luxury brands. Fashion Theory 16(3):273–296
Kunz GI, Glock RE (2005) Apparel manufacturing: sewn product analysis. Prentice Hall, India
Lejeune T (2016) Fast fashion: can it be sustainable? http://source.ethicalfashionforum.com/article/fast-fashion-can-it-be-sustainable. Accessed on Sept 2017
Liz B, Gaynor L (2006) Fast fashion. Emerald Group Publishing, Ltd
Lord PR, Mohamed MH, Ajgaonkar DB (1974) The performance of open-end, twistless, and ring yarns in weft knitted fabrics. Text Res J 44(6):405–414
Mahajan F, Grasso MM (1991) Unpublished research project. University of Texas at Austin
McInerney S (2009). Swap till your fashion footprint drops. http://www.smh.com.au/lifestyle/shopping/swap-till-your-fashion-footprint-drops-20090709-ddx7.html. Accessed on 18 Sep 2017
Mezarciöz SM, Oğulata RT (2011) The use of the Taguchi design of experiment method in optimizing spirality angle of single jersey fabrics. Text Apparel 4:374–380
Ministry of Textiles (2017) Department of industrial policy and promotion. Press Information Bureau, Union Budget 2017–18. https://www.ibef.org/industry/textiles.aspx. Accessed on Sept 2017
Modi D (2013) Upcycling fabric waste in design studio. Thesis submitted to National Institute of Fashion Technology, Mumbai. http://14.139.111.26/jspui/bitstream/1/71/1/upcycling%20fabric%20waste%20in%20design%20studio.pdf. Accessed on Sept 2017
Modi K (2016) Indian apparel sector: government policies drive the growth. www.careratings.com. Accessed on Sept 2017
Nihat Çel K, Çoruh E (2008) Investigation of performance and structural properties of single jersey fabrics made from open-end rotor spun yarns. TEKST KONFEK 4:268–277
Onal L, Candan C (2003) Contribution of fabric characteristics and laundering to shrinkage of weft knitted fabrics text. Res J 73(3):187–191
Puranik P, Jain S (2017) Garment marker planning—a review. Int J Adv Res Educ Technol (IJARET) 4(2):30–33
Rahman M, Haque M (2016) Investigation of fabric wastages in knit t-shirt manufacturing industry in Bangladesh. Int J Res Eng Technol 05(10):212–215
Remy N, Speelman E, Swartz S. Style that's sustainable: a new fast-fashion formula. http://www.mckinsey.com/business-functions/sustainability-and-resource-productivity/our-insights/style-thats-sustainable-a-new-fast-fashion-formula. Accessed on Sept 2017

Sakthivel S, Ramachandran T, Vignesh R, Chandhanu R, Padma Priya J, Vadivel P (2012) Source & effective utilisation of textile waste in Tirupur. Indian Text J http://www.indiantextilejournal.com/articles/FAdetails.asp?id=4236. Accessed on Sept 2017

Savelle BP (1999) Physical testing of textile. Woodhead publishing, Cambridge, England

Sharma IC, Mukhopadhyay D, Agarwal BR (1986) Feasibility of single jersey fabric from open-end spun blended yarn text. Res J 56(4):249–253

Shekhar (2017) Indian textile and garment industry. https://www.linkedin.com/pulse/indian-textile-garment-industry-2017-exporter-manufacturer. Accessed on Sept 2017

Skov L (2002) Hong Kong fashion designers as cultural intermediaries: out of global garment production. Cult Stud 16(4):553–69

Solinger J (1988) Apparel manufacturing handbook-analysis, principles and practice. Columbia Boblin Media Corp.

Solomon M, Rabolt N (2004) Consumer behaviour in fashion. Prentice-Hall, Englewood Cliffs, NJ

Sull D, Turconi S (2008) Fast fashion lessons. Bus Strategy Rev Summer 5–11

Tanvir SI, Mahmood T (2014) Solid waste for knit fabric: quantification and ratio analysis. J Environ Earth Sci 4(12):68–80

Textile Tuesday: ReRoll by Daniel Silverstein. http://bkaccelerator.com/engagesingle/textile-tuesday/. Accessed on Sept 2017

Tokatli N, Kizilgun O (2009) From manufacturing garments for ready to wear to designing collections: evidence from Turkey. Environ Plann 41:146–62

Wicker A (2016) Fast fashion is creating an environmental crisis. Newsweek, US Edition, Tech & science. http://www.newsweek.com/2016/09/09/old-clothes-fashion-waste-crisis-494824.html. Accessed on Sept 2017

Recycled Cotton from Denim Cut Waste

Shanthi Radhakrishnan and V. A. Senthil Kumar

Abstract Denim is associated with the history of success and has endured economic challenges and changes in fashion. Due to its versatility, the connotation it makes as a social statement and acceptance in business meetings and other formal occasions, jeans is considered the top selling 'bottom' in the retail market. According to BernadetteKissane, Euromonitor Apparel and Footwear Analyst, jeans are expected to show a global rate of 3 percent CAGR (compound annual growth rate) by 2020. The manufacturing of jeans involves cutting and sewing the raw material which may be made of cotton, polyester/cotton blend or cotton with elastane. In apparel production, the marker efficiency ranges from 90 to 95% and a high efficiency leads to low wastage to increase profit margins. Whatever be the marker efficiency fabric wastage results as cut part remains which are usually sold in the market for wiping soils and machines. The world today is moving towards zero wastage and sustainable production with the onus on all members in the supply chain, to take responsibility for their business initiatives and consumption. The policy is to use the waste material as a secondary raw material which may be included in the regular production thereby clearing all grounds of wastage. In an effort to help the industry to move towards zero wastage this study was undertaken to utilize the denim cut waste from the apparel industry and convert them into yarn and fabric in conjunction with virgin material. After many efforts recycled cotton fiber was extracted from the denim cut waste by mechanical means and blended with virgin cotton to produce recycled cotton yarns using different blend ratios. As the fiber was colored the dyeing process was eliminated and fabric was produced using recycled blended yarns as weft and 100% cotton white yarn in the warp to resemble the denim fabric. The fabric was tested for physical, mechanical and comfort tests for recommendation as raw material for apparel manufacture. Thus this method of recycling denim cut waste is sustainable and effective in apparel manufacture.

S. Radhakrishnan (✉)
Department of Fashion Technology, Kumaraguru College of Technology, Coimbatore, India
e-mail: shanradkri@gmail.com

V. A. Senthil Kumar
Merchandiser, Knit Fair Inc., Tirupur, India
e-mail: senthilk277@gmail.com

S. S. Muthu (ed.), *Sustainable Innovations in Recycled Textiles*, Textile Science and Clothing Technology, https://doi.org/10.1007/978-981-10-8515-4_3

Keywords Denim cut waste · Recycling · Sustainable manufacture

1 Introduction

1.1 Importance of Material Efficiency in Industrial Production

Strategies which promote sustainability are important tools to design and develop methods for societal progress. Manufacturing companies place their focus on reduction of virgin raw material usage, total raw material utilization, lowest waste generation in production and better waste segregation for reprocessing. These strategies may be long term or short term in focus on material efficiency and solutions for waste management. To mention a few are Cleaner Production, Waste minimization, Eco design, Best practices, Closed loop, Reverse Logistics, Industrial Ecology, Product stewardship, Environmental Management, Eco mapping, Waste hierarchy, Material flow cost accounting and resource efficiency.

Cleaner Production involves the continuous application of a strategy towards products, processes and services with a concern for the environment. The strategy can have a three focus approach namely reduction of resource consumption and waste at source; reuse and recycling; product modification. Waste reduction at source calls for strict administrative preventions (water and energy losses), process control (check on pressure, time, temperature and pH), material substitution (use of water soluble PVA instead of starch for sizing; use of plasma treatment instead of chlorination for shrink proofing) (Toprak 2017) and equipment modification (optimization of processes, setting engine speed).

Reuse and recycling contributes to the utilization of waste in a productive and economically viable way. The American & Efird (A&E), USA has diverted 20,000 lb of waste thread each month from ending into landfills through efficient recycling programs. In Tamilnadu, India, Vardhman Yarns and Thread Limited have entered into a joint venture with A&E, USA and have a zero liquid discharge ETP which uses 93% of the of the dye house water while the remaining 7% waste is evaporated through air, mechanical and solar evaporators and reused (Vardhman 2012; Leonas 2017). Some of the organizations worth mentioning in the field of textile recycling are The Smith Family Commercial Enterprise, Australia; ZAO 'ZavodTver'Mash, Russia; Tie and Dye technique (Adire) on used material by the Yoruba Women of South west Nigeria; GOONJ project, Khisco Group, New Delhi, India and Solid Waste District La Porte County, USA (Roznev et al. 2017).

The processes involved in making a product may have many methods which may cause pollution. One of the basic aspects of cleaner production is to reduce pollution and may require a change in product specification and the functionalities it will perform. Product modification may require reduction in weight or thickness of the product, change in packaging and optimizing the minimum packaging required to protect the product are some of the methods that allows for recycling or reuse of the product (Toprak 2017). An example of product modification is the design concept of halfway products. The creativity of the consumer is enhanced by the provision of product kits suitable for quick disassembly and reassembly. The modular structure of the garment allows the consumer to mix and match or upgrade certain parts or sections to create a new look. Further the parts that are soiled can be detachable and removed and washed separately (Papanek 1995; Faud-Lake 2009; Niinimäki and Hassi 2011; Fletcher 2008).

With regard to waste minimization a few case studies have been reviewed. The waste minimization procedure formulated by the United States Environmental Protection Agency (US EPA) was tested in a piece dyeing mill by producing mass and energy balances; steam distribution, steam utilization and condensate system was analyzed and reviewed and the waste minimization scheme was proposed. The implementation reduced waste water (less 21%), dye consumption (less 24%), textile auxiliaries and chemicals (less 14%), steam usage (less 25%) along with flu gas and green house gases; the payback period for the proposed scheme was less than one year (Moore and Ausley 2004). A study on the optimization of desizing parameters in the textile industry (ratio of desizing agent to Fabric = 20 g/g; temperature 80 °C; time 7 s) showed that 89% of the sizing agent was eliminated and the effective use of rinse water, chemicals and energy resulted in cost reduction and decrease in waste water generation; the production capacity increased from 30 to 34.4 m/min (Tanapongpipat et al. 2008). The best textile recycling capacities are in Germany and they work in collaboration with ten partners for developing innovative products for specific uses like insulation materials for acoustics, recycled polyester fabrics for uniforms and heavy duty canvas material from recycled fibers (Ahmad et al. 2016; Gulich 2006).

Eco designs are those products that have great concern for environment, health and safety issues during manufacture, use and after the use of the product. When environmental concerns are greater the quality of the product is enhanced in terms of function, dependability, strength and reparability. DfE is a term which is commonly used, as synonym with 'Design for Recycling' or 'Design for Disassembly'. Specialists in eco design should have a flair for designing products that are classic, long lasting and having the possibility of repair, reuse, disassembly and recycling. Eco designs also calls for optimization of the existing production processes, finding alternatives to raw materials and production processes and the development of new product concepts (Crul and Diehl 2006). EVTEK Institute of Art and Design, Finland, had conducted a research study on customers' interest in eco products under the Natural Dyes Product Development Project 2000–2003. It was found that a minority of consumers these days look out for eco design and eco textiles which are usually linked with higher prices. Though in general consumers look out for the 'green' label, factors like price, quality and aesthetics play an important role in deciding the purchase of a product (Niinimäki 2006).

Best practices are methods followed in the industry with the target of reducing waste and utilizing material to its fullest efficiency. These practices are chosen from the general options available for waste minimization to suit the needs and requirement of the said industry or organization. Usually the options are classified into two categories namely general waste minimization options which gives suggestions for reducing the use of available resources and the second set of options are given in terms of each textile process. Reports and guides are available for industries to select the best practices for waste minimization and material efficiency e.g. Best Management Practices for Pollution Prevention in the Textile Industry by USEPA (EPA 1996), Environmental Technology Best Practice Program—Textiles Workbook jointly accomplished by the Department of Trade and Industry and the Department of Environment, UK (ETBPP 1999), Industry and Environment, United Nations Environment Program (UNEP 1996), Cleaner Technology Transfer to the Polish Textile Industry, DEPA, Denmark (Wenzel et al. 1999). Many case studies have also been mentioned to highlight the efficiency of the material chosen for the industrial processes which could be an alternative or substitution to the conventional raw material or auxiliaries. Use of Catalase enzyme formulation instead of the traditional reducing agent in the bleaching process (The Skjern Tricotage—Farveri textile mill, Denmark), reducing sulphate in the effluent by chemical substitution: sodium chloride (Quimica y textiles Proquindus SACI, Chile), reducing COD in effluent by replacing starch size with water soluble starch and PVA (The Misr Spinning and Weaving Company, Egypt), reducing COD in effluent by replacing soap in scouring with anionic/nonionic detergents (Scottish Finishing Company, Scotland) (Barclay and Buckley 2000) are some of the case studies which focus on the appropriate selection and use of raw materials for industrial processes.

In the closed loop system it may be understood that the components which constitute a product, at the end of its functional life, should be split down and reconstituted as a part or whole into new products. Circle Economy, in 2016, aligned with Recover, ReBlend, G-Star RAW, ReShare and Wieland textiles to venture into high value recycling. The key learnings from these joint ventures showed that the new denim recycled fabrics produced fetched a premium price of 12.5% to virgin cotton denim fabrics with only 12% recycled content; 7 tons of post consumer garments were converted to 6 tons of 100% recycled yarns with a reduction of 62% water consumption, 33% energy consumption 18% green house gas emissions; several tons of old Dutch navy and army uniforms were used to produce recycled yarns for the production of humanitarian aid blankets (Circle Economy 2017).

Any industrial processes can be assessed by studying the material and energy flows of the industrial system/production which is termed as IE or Industrial Ecology. Taking the case of Tirupur town in Tamilnadu which produces knitwear for export, the study conducted a resource flow analysis and the outcome highlighted that materials for reuse accounted to 2430 tons/year of plastics, 25,532 tons/year of fabric, 20 tons/year of metal and unused resources accounted to 87,500 thousand liters/day and 54,492 tons/year solid waste (Anonymous 2017). The resource flow analysis also enlightens that the industrialists are collectively spending US $7 million annually for buying water along with the burden of maintenance cost of effluent treatment plant. On the basis of this study, a system has been set up which uses the waste heat from the boilers to recycle water at a relatively low cost; the solid waste garbage has large quantities of paper and textiles having high calorific value, which could be used to replace the firewood used in Tirupur for industrial production. There is a development of setting a central steam plant to replace the firewood usage thereby reducing the rapid deforestation around the place. Similar industrial ecology studies have also been undertaken for the Leather Industry in Tamil Nadu to solve many industrial problems and to reuse the waste to get better results (Lew 2017).

Product Stewardship and Extended Producer Responsibility are used interchangeably to represent the people or personnel responsible for waste management. All involved in the creation, use and the end-of-life management of a product must take responsibility to safeguard the environment from the impact created by the product at all stages of the life cycle. The major responsibility lies in the hands of the producer, followed by the retailer and consumer. The responsibility of collection, transportation and management of product waste has shifted from the Government to the citizens who play a major role in generating the waste. Australian Post has bagged the APC Award for Outstanding Achievement in Packaging Stewardship in 2016, recognizing its key role in the recovery of a range of products which include APC's cross industry coffee cup recycling, collection of used cartridges and mobiles in conjunction with Planet Ark and Mobile Muster, collection of used button and coin cell batteries through the Australian Battery Recycling Initiative, joined hands with Terracycle for recycling coffee pods, tooth brushes and cleaning articles and its involvement in Tyre Stewardship Australia (APCO 2017). The selection of material and the designing of the product stand as a major tool in impacting the environment.

Eco mapping is a preliminary step in Environmental Management systems. It gives visual representations by a multi layer set of graphical information of the environmental hotspots on the industry's layout. Environmental problems, material flows and utilization, workers opinion and work processes are well understood by the environmental teams to plan for futuristic needs. Environmental Management systems (ISO 14001–14009) help industries to focus on the environmental standards and ensure a clean, health and safe world. The implementation of the EMS system ISO 14001 in Alps Industries Limited, North India, has resulted in savings—reuse of dye liquor in indigo dyeing for an additional ten cycles after replenishment with cost benefit of Rs. 2.00 per kg and reduction of effluent load; reduction of air pollution by humidification and water jets; reuse of selvedge waste for embellishment; recycling of spinning and weaving waste for yarn development (APCO 2017; Joshi 2001).

Waste hierarchy by EU indicates prevention of waste as the most important step in the ladder of waste management. Procurement of raw materials and disposal of waste materials is expensive these days and recycling practices are the need of the hour. A combination of polyester thread from recycled PET bottles and shredded fibre from apparel cut waste by EcoGear; recycled rubber from motorcycle and bicycle tubes for the manufacture of fashion apparel; composites from recycled polypropylene, silk and cotton waste produced in Turkey; North Carolina State University and Burlington Industries have jointly produced reclaimed denim with 50% reclaimed denim yarn; Esleeck, US has produced Blue Jean Bond Paper from cotton rags for legal documents and correspondence (Bhatia et al. 2014; Taşdemır et al. 2007; Blackburn 2015; Crighton 1993).

Material flow cost accounting helps an organization in tracing waste as materials, energy losses and emissions through the processes involved. GUNZE Limited, Japan specializing in inner wear had conducted a study to analyze the MFCA and the process selected was weaving of fabric to the finished garment. All the materials used in the process sequence came under the purview of the MFCA calculation and the material losses in terms of volume and cost were tracked and estimated. The material flow cost matrix showed that material cost contributed to 45.1%, energy cost 2.9%, system cost 51.4% and waste management cost 0.6% (Schmidt and Nakajima 2013). The MFCA analysis highlighted that it was necessary to set an appropriate standard for new materials used in the product development and design phase. This analysis tool was very suitable for resource efficiency in apparel industries where constant use of new materials and changing fashion was evident.

1.2 Case Study on Recycling of Denim Cut Waste

From the above it can be understood that material efficiency is the most important factor in product development followed by the systems used for the conversion of raw material to the finished product. The manufacturer's role does not end there but is held responsible for the use and end of life management of the product. Products designed with functional raw materials which also fulfil the role of reuse and recycle into new products or as secondary raw materials are required for the ultimate goal of sustainability. To explain the importance of the use of recycled raw materials, a case study has been taken as an example where the denim cut waste from M/s. KG Denim, Coimbatore has been recycled to form yarns and woven into fabric using plain weave with the recycled blue yarn in the weft and the virgin white cotton yarn in warp direction. The developed fabric resembled denim but was lighter in weight and thickness and made into a product (skirt) suitable for children's apparel.

2 Objectives

The aim of the case study is as follows:

- Sourcing of denim cut waste
- Extraction of fibers from the sourced raw material
- Conversion of recycled fibers into yarn and fabric
- Product development from fabric developed from recycled yarns.

Table 1 Properties of the virgin cotton fiber and recycled cotton fiber used for the study

Parameters	Strength (grams in tex)	Elongation (%)	Length (mm)	Fineness (microgram/inch)
Virgin cotton fiber	34	6.3	26	3.2
Recycled cotton fiber	28.25	6	20	3.6

3 Materials and Methods

3.1 Materials

Textile waste may be classified as pre consumer waste or post consumer waste. Waste produced during manufacturing and becomes waste material before it reaches the customer is pre consumer waste (Leonas 2017; Lau 2015; Fiber2fashion 2017; Well made Clothes 2017; Hickman 2010). In apparel production the primary motto is zero waste (Parellax Limited 2017; Jenkin 2015; Claudio 2007; Nayak and Padhye 2013) but while marking the pattern components a small percentage is allowed to enable knives or lasers to move around the marked areas and enable cutting of the fabric lays. Secondly marking losses occur when there are gaps or non-usable areas between the pattern pieces in the marker. While making the lays for cutting other losses occur which include end of ply and piece losses, edge losses, splicing losses, remnant losses and ticket length losses which may range from 0.5 to 5% (ITJ 2008). Fabric wastes from the cutting department serve as an opportunity to recycle to create new products or intermediate products. The raw material used for the study is pre consumer fabric (denim) cut waste sourced from the cutting department of the jean manufacturing company.

Any recycled material has to go through a process of recycling be it thermal, chemical, mechanical in nature before it is converted into a raw material or intermediate material to enter into production (Wang 2010; Al-Salem et al. 2009; Hawley 2006). These processes lower the strength properties of the recycled material and hence there is a tendency to blend virgin material to compensate the losses (Necef et al. 2013; Jankauskaite et al. 2008; Rowe 2000). Virgin cotton was selected and blended with the recycled fibers to produce strong yarns suitable for apparels. The properties of the virgin cotton and the recycled cotton fibers are given in Table 1.

3.2 Methods

The methodology used for the study is given in Fig. 1.

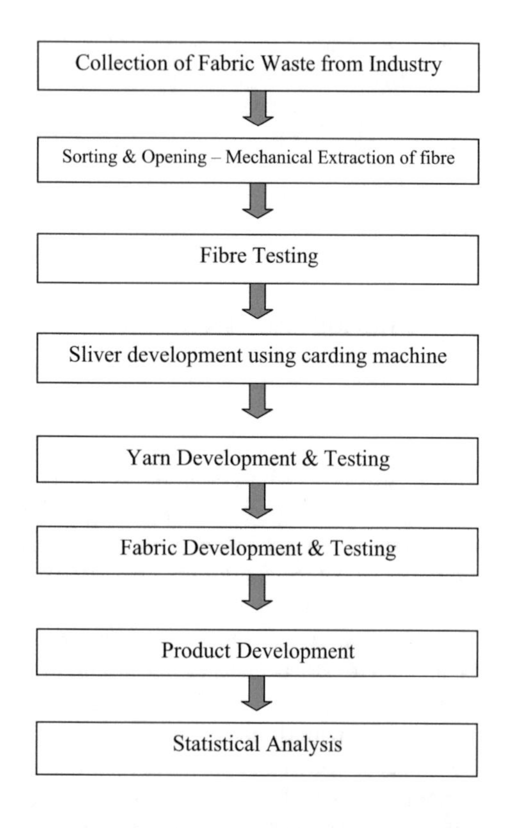

Fig. 1 Experimental design of the study

- **Collection of fabric waste**: Fabric waste in the cutting department of the apparel industry is due to improper color shade matching and inefficient marker planning. The raw material denim cut waste was collected from KG Denim Limited, Coimbatore which specializes in jean production under brand name 'Trigger' (Denim 2017).
- **Sorting and opening—Mechanical Extraction of fiber**: The denim fabric cut waste has been converted to fiber mechanically using the hard waste opening machine. The input raw material was fed manually through a wooden conveyor belt of the first opener to be automatically transferred to bottom of the spiked roller. This roller is responsible for opening the denim cut waste into fiber form. By controlling the speed of the feed roller and spiked roller, the maximum fiber length can be obtained. The diameter of spiked roller is 26" and the feed roller is 9". The denim cut waste used for the project and the hard waste opening machine is given in Figs. 2 and 3 respectively. The output of the machine as recycled fibers is shown in Fig. 4. The flowchart for opening the denim waste into fiber form is represented in Fig. 5.
- **Fiber Testing**: The fibers obtained from denim cut waste were tested for Fiber Strength and Elongation using the Stelometer [ASTM D1445], Fiber Length using Baer Sorter [ASTM D1447], Fiber Fineness using the Digital Fiber fineness Tester

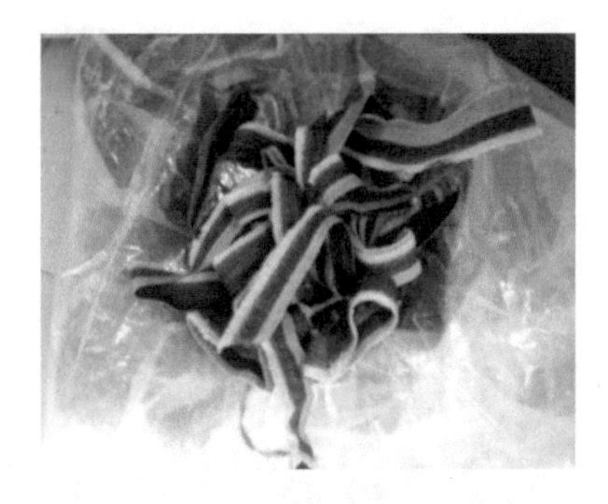

Fig. 2 Fabric cut waste [Denim]

Fig. 3 Hard waste opening machine

Fig. 4 Denim cut waste into fiber form

[ASTM D 1448], Fiber Composition Test and Burning Test. The results of the Fiber Strength, Elongation, Fiber length and fineness are given in Table 1.

Denim fabric may be composed of a combination of fibers for functional reasons and durability. Spandex fibers contribute to the strength and stretch factor of the denim fabrics. To identify the presence of spandex the chemical test for spandex was carried out. Spandex is soluble in Dimethyl Formamide. Cotton fiber is not soluble in Dimethyl Formamide. 100 g (w1) of denim fiber is added to 30 ml of 100% concentration Dimethyl Formamide. The water bath was heated to 90 °C and the contents of the bowl was kept on the hot water for 10 min. After 10 min the fiber was extracted from the chemical and kept in hot oven for 30 min. After full drying

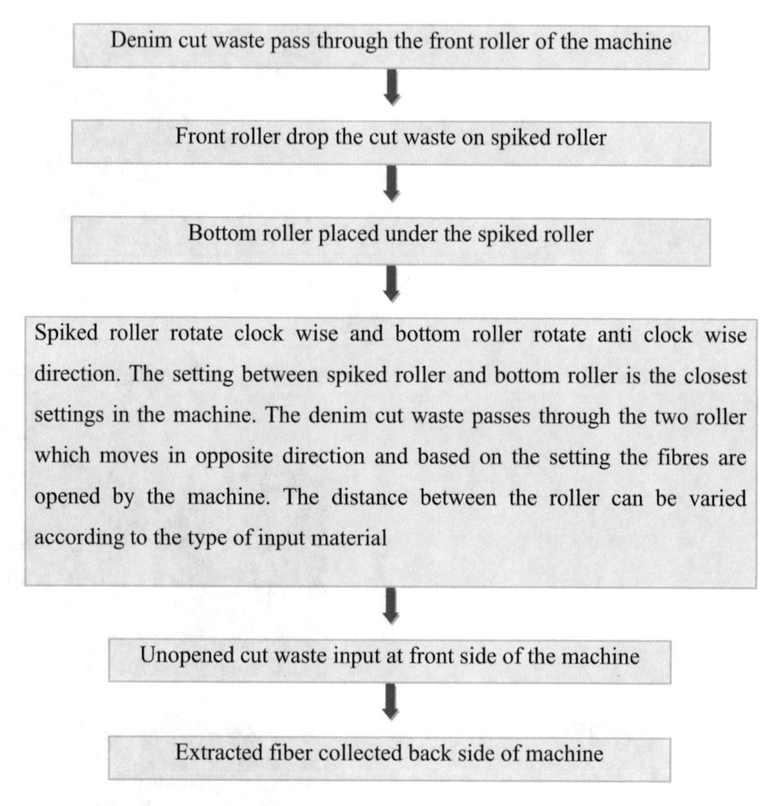

Fig. 5 Flow chart for opening the denim cut waste

the weight of the fiber (w2) was taken. The difference between the two weights gave the weight of Spandex.

$$\text{Weight of spandex} = \text{w1} - \text{w2}$$

Ten samples were taken at random and were tested. The mean value showed that there was no change of weight indicating the absence of spandex.

The confirmatory burning test was done to identify the presence of cotton. Denim cut waste fabric warp and weft yarns were unravelled. Denim cut waste warp and weft yarn was shown on fire flame. It burned quickly with yellow flame and when the yarns were removed from flame it continued to burn rapidly with a afterglow. The odour was like burning paper with grey smoke and grey color ash. Ten samples were selected at random and tested. The test results also confirmed the absence of polyester as no melting and bead formation was seen.

Fig. 6 Conversion of carded web into sliver

Table 2 Machine settings for sample carding machine

Feed roller speed [RPM]	0.533
Cylinder speed	692
Doffer speed (for lap)	3.4
Doffer speed (for sliver)	3.8

- **Sliver development using carding machine**

The recycled fibres obtained from the denim cut waste was sent to the Trytex Sample Carding Machine for sliver development as seen in Fig. 6. Carding machine roller speed settings given Table 2. The mechanisms for fibre transfer between carding roller surfaces, will provide much needed insight into general fibre dynamics during the web forming process, machine design and process control optimization, and the formation of the sliver (Ichim and Sava 2016; Sharma and Goel 2017).

In this study four yarns were developed as mentioned below.

- 100% recycled cotton fibers [100 RCYa]
- recycled cotton fiber: virgin cotton fiber = 80:20 [80/20 RCYb]
- recycled cotton fiber: virgin cotton fiber = 60:40 [60/40 RCYc]
- recycled cotton fiber: virgin cotton fiber = 40:60 [40/60 RCYd].

The ratio of the fibers was maintained by weight and was mixed during the preparation of the carding process by manual method.

- **Yarn Development and Testing**: Open end spinning was chosen for yarn development. In open end spinning, the sliver is opened into individual fibers by means of an opening roller followed by the transportation of the fiber by air. A change in speed is devised locally at a predetermined position that drops the number of fibers in the cross section enabling twist by the rotation of the yarn end leading to a continuous strand of yarn which is wound to a rotating bobbin (NPTEL 2013; Barella et al. 1973; Abou-Nassif 2014; Yu 1999; Textile School 2017) as shown

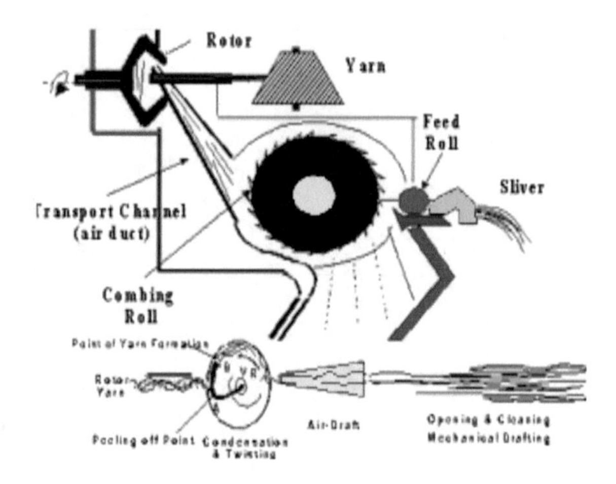

Fig. 7 Working principle of open end spinning

Table 3 Machine settings for sample open end spinning machine

Yarn count	20 Ne
Feed hank	0.25
Twist direction	Z
Open roller speed	30,000 rpm
Yarn length	5000 m

Fig. 8 Recycled OE yarn development

in Fig. 7. This method of spinning is suitable when the raw material fiber length is lower and the resultant yarns are coarse (Bagwan and Patil 2016; Online Textile Academy 2017).

The Trytex Sample Open End Spinning Machine was used for producing 20's OE yarn from the recycled denim fibers. The rotor diameter was 41 mm and the rotor speed was maintained at 30,000 rpm and the TPI was 30, for the development of all the yarn samples. The machine settings for the OE spinning is given in Table 3 and the yarn development is shown in Fig. 8.

Table 4 Body measurements for Skirt

S. No	Measuring points	Measurements
1	Full length	14.5″ (38 cm)
2	Waist	24″ (60 cm)
3	Seat	30″ (75 cm)
4	Belt width	1″ (2.54 cm)

Yarn Testing: The yarns developed were tested for Yarn Count (ASTM D 3375), Single thread strength and Elongation (ASTM D1578), Lea strength test (ASTM D1578), Yarn evenness test and Yarn appearance (ASTM D2255)

- **Fabric Development and Testing**: Five different fabrics have been developed from virgin cotton yarns and recycled yarns; 100% virgin cotton yarns are being used as the warp and the developed recycled yarns will be used as the weft yarns. The fabrics produced in the power loom are given below.
  - Warp: 100% virgin cotton OE yarn Weft: 100% Recycled OE yarn [RCFa]
  - Warp: 100% virgin cotton OE yarn Weft: 80/20 Recycled OE yarn [RCFb]
  - Warp: 100% virgin cotton OE yarn Weft: 60/40 Recycled OE yarn [RCFc]
  - Warp: 100% virgin cotton OE yarn Weft: 40/60 Recycled OE yarn [RCFa]
  - Warp: 100% virgin cotton OE yarn Weft: 100% Virgin cotton OE yarn [VCF]

  The tests undertaken for the developed fabrics are as follows

- Physical Tests: EPI, PPI [ASTM D 3775], Fabric weight [ASTM D3776], Fabric thickness [ASTM D 1777]
- Mechanical Tests: Tensile strength and elongation [ASTM D 5034], Stiffness test [ASTM D 1388], Fabric crease resistance [ASTM D 1296], Fabric abrasion Resistance [ASTM D 4966], Fabric drape [ASTM D 3691], Pilling test [ASTM D 3512]
- Comfort Tests: Air permeability test [ASTM D 737], Wickability test [ASTM D 2692], Thermal conductivity test [ASTM D 6343], Water vapor permeability test [ASTM standard E 96], Spray rating test, Sinking test, Color fastness, Antimicrobial test [AATCC 100]
- **Product Development**: The developed material was constructed into an Aline Skirt and worn on a dummy and then on a live model to find out the appearance and comfort. The measurements for the skirt are given in Table 4 and Fig. 9. Patterns were drafted and the fabric was cut into component parts to be constructed into a skirt.
- **Statistical Analysis**
  Statistical analysis involves scrutinizing a set of data to enable the researcher to draw conclusions. Statistical analysis includes collecting and analyzing all the samples in a data set from which the samples are drawn (Rouse 2017). Readings were recorded for each of the tests identified for fiber, yarn, fabric properties. Use of the statistics to analyze control and four experimental fabrics was carried out. Means were used to analyze the fabric properties. Inferential statistics (Analysis

Fig. 9 A line skirt (Talyq 2017)

of Variance) was employed in testing the data since they are measures used in measuring differences, and the purpose of this study was to establish if any difference existed between the control and the experimental groups involved in the study. Virgin cotton as use as control and recycled cotton in the experimental groups. The variables were analyzed for differences between the group and among the group. A small *P*-value results in greater F-values. 1 and 5% significance levels were used for analysis. Analysis of variance (ANOVA) is a collection of statistical models used to analyze the differences among group means and their associated procedures (such as "variation" among and between groups) (Lane 2017).

- **Cost analysis and saving analysis**
  A cost analysis is detailed outline of the new product development cost. New product development requires initial monetary investments to buy raw material. The cost analysis includes cost of fibre extraction, yarn development, fabric development and desizing. Cost analysis is used to predict bulk quantity of product development cost with different ratios of fibres.
  Saving analysis or benefit analysis is comparing the saving of expenditure for new product or commercial product. Saving analysis plays an import role in the recycled product development. Saving analysis include cost of raw material, conversion to yarn and fabric, dyeing and finishing. While comparing fabric with different blend ratio's product development cost 100% recycling given more saving.

4 Results and Discussion

The results of the study are discussed under the following heads:

4.1 Yarn test results
4.2 Fabric test results
4.3 Cost and saving analysis.

The results of the virgin cotton yarn and the recycled yarn tests are given in Table 5. The 100% virgin cotton yarns have been taken as the control and the yarns developed from the recycled yarns have been compared to show the difference between the two in all parameters.

From Table 5, it may be understood that there is a reduction in the strength (single yarn and lea strength) and elongation in all the recycled yarns when compared to the virgin cotton yarn. As the percentage of virgin cotton yarn increases there is a gradual betterment in strength and elongation which shows that the percentage of virgin cotton yarn added can be based on the application for which the yarn will be used. The maximum loss in strength and elongation is seen in the recycled yarns made of 100% recycled fiber from denim cut waste where no virgin cotton has been added. This may be attributed to the mechanical recycling process where the fibers are opened by the teeth or spikes of the rollers leading to fibers of lower length and strength. Further these fibers are subjected to a spinning process for the second time after the mechanical opening of the denim cut waste (Telli and Babaarslan 2017). In all the three tests single yarn strength, elongation and lea strength the F value is high indicating that the p value is lower than 0.05 showing that there is difference between the means in the ANOVA test results (Frost 2016).

Table 6 shows the physical properties of the recycled cotton yarns in comparison with the virgin cotton yarn in terms of thin places, thick places, neps and hariness. This shows the quality of the yarn produced with the recycled cotton fibers obtained from the denim cut waste.

Table 6 indicates that number of thin place, thick place and neps have increased when compared with the 100% virgin cotton yarn. As the virgin cotton fiber content increases the there is a gradual decrease of thin, thick places, neps and hairiness. With regard to hairiness there is increase in values showing deterioration in the yarn quality of recycled yarns. The mechanical process involved in recycling produces shorter fiber length. While spinning the short length of fibers cause irregularity like thin places, thick places and neps the protrusion cause increase in hairiness. Further the rotor speed and the opening roller speed have a significant effect on the thin places, thick places, neps and hairiness of the yarns (Wanassi et al. 2015). Statistical analysis shows that there is significant difference at 5% level in all the parameters studied.

4.1 Fabric Test Results

The five fabrics developed were tested for count, weight, thickness and ends per inch and picks per inch. Since the EPI and PPI were kept equal during the weaving process there was not much changes in all the samples. In the case of yarn count there was very meager differences, but in the weight and thickness of the samples the ANOVA results showed that there was significant difference between the means at 0.05% level of significance.

Table 7 highlights that the fabrics made from recycled yarns showed greater weight and thickness when compared to the fabrics made from virgin cotton. This may be due to the irregularities in the fiber taken from the denim waste that was sourced

Table 5 Properties of the virgin cotton and virgin cotton/recycled cotton blended yarns

S. No	Particulars	Control 100% VCY	100 RCYa	%Gain or loss over control	80/20 RCYb	%Gain or loss over control	60/40 RCYc	%Gain or loss over control	40/60 RCYd	%Gain or loss over control	F Value
1	Single yarn strength (kg/sq.cm)	336.2	300.9	−10.5	310.5	−7.64	325	−3.33	328.4	−2.32	15,058.74[a]
2	Elongation %	13.06	10.2	−21.89	11.03	−15.54	11.88	−9.04	11.2	−14.24	19.31[a]
3	Lea strength (lbs)	63	59	−6.35	58	−7.94	60	−4.76	62	−1.59	462.43[a]

[a]Significant at 5% level

Table 6 Properties of virgin cotton and virgin cotton/recycled cotton blended yarns

S. No	Particulars	Control 100VCY	100 RCYa	%Gain or loss over control	80/20 RCYb	%Gain or loss over control	60/40 RCYc	%Gain or loss over control	40/60 RCYd	%Gain or loss over control	F Value
1	Thin place (%)	33	397.5	1104.55	255	672.73	100	203.03	85	157.58	115.42[a]
2	Thick place (%)	75	692.66	823.54	460	513.33	373	397.33	119.5	59.33	462.12[a]
3	Neps (%)	95	508	434.737	508	434.74	360	278.95	385	305.26	127.56[a]
4	Hairiness (%)	5.26	6.17	17.300	6.02	14.45	5.47	4.00	5.37	2.09	404[a]

[a]Significant at 5% level

from the apparel industry. Further the fiber test results show that the fiber length is shorter than the virgin cotton which causes protrusions and variations in thickness. The fibers constituting the yarn must be of medium to good quality to get good results and open end spinning also forms coarser yarns when compared to ring spinning. It has also been reported that open end spun yarns have lower fiber packing density that makes the yarn more bulky in nature as open end spinning principle is based on the assembly of twisted fibers (Hearle et al. 2008). The weight and thickness of the fabrics gradually decreases as the percentage of virgin cotton increased (Wanassi 2015; Sultan and El-Hawary 1974).

While examining the mechanical properties of the fabrics in Table 8, there is a reduction in the bending length in all the samples when compared to virgin cotton control samples in the warp and weft direction. When the bulk of the samples made from recycled fibers increases the weight and thickness have a bearing on the bending length of the samples. Similar trends were noticed in the bending modulus of the samples. In the case of flexural rigidity there was an increase in the recycled samples showing that the fabrics made from recycled fibers were stiffer and more rigid when compared to the virgin cotton control samples. The drape coefficient results also confirmed the same. Crease recovery results showed an increase in the warp and weft directions indicating that the recovery of the fabrics is slower than the control virgin cotton samples. The colour fastness to rubbing was done just to measure the colour fastness of the sample. Normally denim fabric is dyed to slowly fade and hence this quality will be persistent in the fabric.

Table 9 shows a decrease in air permeability in all the recycled samples as recycled yarns are thicker then the virgin cotton yarn. The pores between the yarn in recycled fabric are comparatively low when compared to virgin cotton yarn fabric. A decrease in spaces between the yarn indicate the decrease in water vapour permeability and thermal conductivity as there is low movement for water and heat in and out of the fabric.

Table 10 shows a mixed trend of responses to the comfort tests performed. In the case of wickability there is slower wicking property in the 100% recycled fabrics but as the virgin cotton composition increases there is better wickability in the warp and weft direction. In the case of spray rating there is not much of a difference between the samples. In the case of sinking test the time taken for sinking is greater in all the samples when compared to the virgin cotton fabrics. This may be due to the difference in the weight and thickness of the recycled fabrics samples and it takes more time to sink to the bottom of the beaker. The F values for wickability and sinking tests show that there is significant difference between samples at 5% levels.

Table 11 and Figs. 10, 11 and 12 reveals the absence of bacteria as the zone of inhibition is almost zero (Kaur et al. 2012; Bhalodia and Shukla 2011; Reller et al. 2009). Many consumers may feel that the apparels made from recycled materials are susceptible to bacterial growth and infection, but the yarns go through all the processes involved in fabric processing and hence they are free from these notions. Since this was a study done on a small scale care was taken at all stages to keep the raw material clean and subject it to safe processing. Further the cut waste has been stored and sold in a clean environment at the place of purchase and hence the chances

Table 7 Physical properties of fabrics from 100% virgin cotton and virgin cotton/recycled cotton blended fabrics

S. No	Particulars	Control 100% VCF	100 RCFa	%Gain or loss over control	80/20 RCFb	%Gain or loss over control	60/40 RCFc	%Gain or loss over control	40/60 RCFd	%Gain or loss over control	F Value
1	Yarn count (Ne)	20.01	20.09	0.39	20.03	0.09	20.03	0.09	20.02	0.04	
2	Fabric weight (gsm)	157	182	15.92	167	6.37	159	1.27	163	3.82	2778.75[a]
3	Thickness (mm)	0.56	0.62	10.67	0.6	6.76	0.57	2.84	0.57	1.60	6.91[a]
4	EPI	56	56	0	56	0	56	0	56	0	
5	PPI	60	60	0	60	0	60	0	60	0	

[a]Significant at 5% level

Table 8 Mechanical properties of fabrics from 100% virgin cotton and virgin cotton/recycled cotton blended fabrics

S. No	Particulars	Control 100 VCF	100 RCFa	%Gain or loss over control	80/20 RCFb	%Gain or loss over control	60/40 RCFc	%Gain or loss over control	40/60 RCFd	%Gain or loss over control	F Value
1	Bending Length (cm) (Warp)	2	1.7	−10	1.85	10	1.9	−5	2	0	15.86[a]
	Weft	2	1.6	−20	1.8	−12.5	1.7	−6.25	1.7	−6.25	14.37[a]
2	Flexural Rigidity (mg.cm)	44.64	60.58	35.70	76.01	70.27	78.81	18.30	80.67	6.78	3724.19[a]
3	Bending Modulus (kg/cm^2)	0.003	0.001	−63.33	0.0015	−50	0.0018	−63.33	0.002	−67.33	302.70
4	Fabric Drape (%)	47.75	39.03	18.26	43.9	8.06	46.8	1.98	47.03	1.50	277.47[a]
5	Crease Recovery (warp)(°)	100	102	2	102	2	105	5	106	6	6.11[a]
	Weft	100	101	1	105	5	104	4	105	5	11.01[a]

[a]Significant at 5% level

Table 9 Comfort properties of fabric from 100% virgin cotton and virgin cotton/recycled cotton blended fabrics

S. No	Particulars	Control 100 VCF	100 RCFa	%Gain or loss over control	80/20 RCFb	%Gain or loss over control	60/40 RCFc	%Gain or loss over control	40/60 RCFd	%Gain or loss over control	F Value
1	Air Permeability ($cc/sec/cm^2$)	191	185.68	−2.78	188.15	−1.49	190.74	−0.1	190.2	−0.42	17.7[a]
2	Water Vapour Permeability ($g/m^2/day$)	8.54	7.15	−16.31	7.61	−10.90	7.82	−8.42	8.02	−6.08	1449.27[a]
3	Thermal Conductivity (w/m/k)	0.04	0.01	−61.90	0.019	−54.28	0.02	−42.38	0.039	−7.1429	1167.71[a]
4	Color fastness to rubbing	0	3	0	3.4	0	4.5	0	4.5	0	

[a]Significant at 5% level

Table 10 Comfort properties of fabrics from 100% virgin cotton and virgin cotton/recycled cotton blended fabrics

S. No	Particulars	Control 100 VCF	100 RCFa	%Gain or loss over control	80/20 RCFb	%Gain or loss over control	60/40 RCFc	%Gain or loss over control	40/60 RCFd	%Gain or loss over control	F Value
1	Wickability (Warp) (sec/cm)	605	654	8.09	545	−9.91	481	−20.49	429	−29.09	1077[a]
	Weft	425	450	5.88	420	−1.17	405	−4.70	397	−6.58	61.41[a]
2	Spray rating test	50	50	0	50	0	50	0	50	0	
3	Sinking test	12	16	33.33	15.25	27.08	14.5	20.83	14	16.66	75.37[a]

[a]Significant at 5% level

Table 11 Antibacterial test results of 100% recycled fabric

S. No		Sample	Antibacterial activity (Zone of inhibition in mm)	
			E. Coli	*S. Aureus*
1	Control fabric	100% Denim RC sample	0	0
2		Cotton 100% fabric	0	0

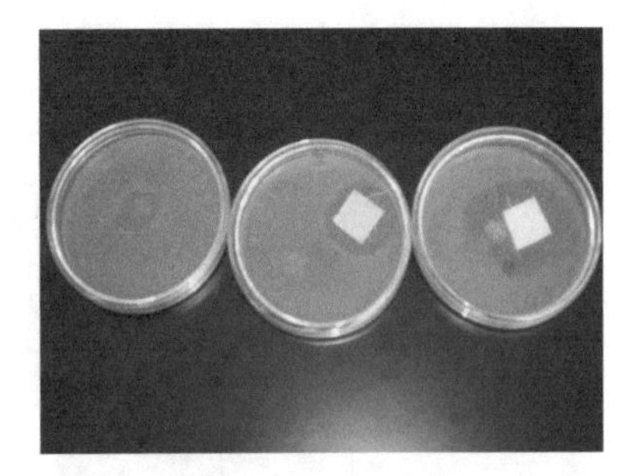

Fig. 10 100% cotton fabric *E. Coli* and *S. Aureus*

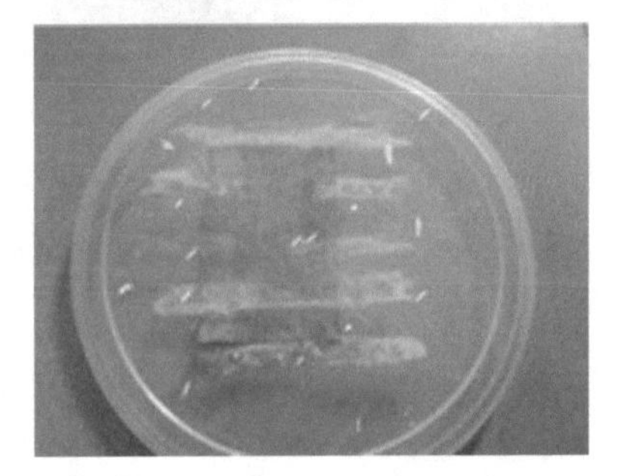

Fig. 11 *E. Coli*—100 Denim recycled fabric

of deterioration are less. There are cases where the fabric waste in kept in unhygienic conditions, in such cases more care must be taken during the development of the fiber, yarn and fabric. This test shows that the developed fabric is equivalent to the virgin cotton fabric in terms of antibacterial test.

4.2 *Cost and Saving Analysis*

The cost and saving analysis is given in Tables 12, 13, 14 and Fig. 13. The raw material cost for recycled cotton includes cost of raw material and the cost for mechanical fiber extraction. When it is blended with virgin cotton (80/20, 60/40 and 40/60), the virgin cotton cost is also added. The virgin fiber cost is the commercial cost incurred during the purchase of the fiber. The cost for developing the recycled yarn includes open-

Fig. 12 *S. Aureus*—100% Denim recycled fabric

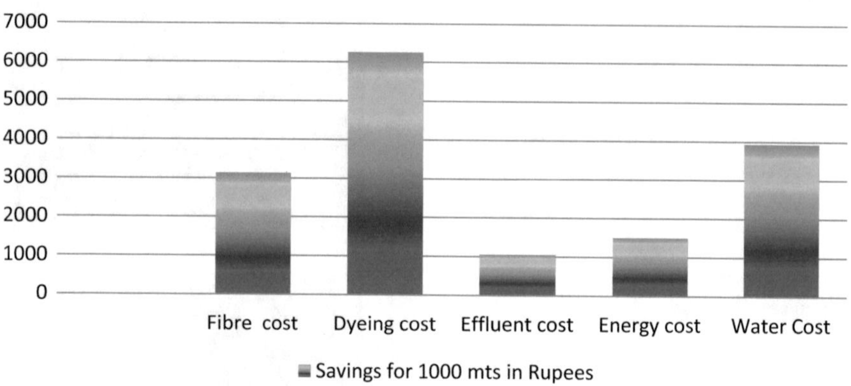

Fig. 13 Savings in cost—recycled cotton fabric vs virgin cotton fabric development

end spinning charges paid at KCT-TIFAC Core where the sample yarn development was undertaken. The charges for spinning will be lesser when it is carried out in an industrial scale where bulk quantities will be used. Table 12 shows the cost in rupees for procuring the raw material, mechanical extraction of the fiber and the yarn development charges. The cost has been calculated for 1000 m of fabric for the sake of convenience. The requirement for the production of 1000 m of fabric is 125 kg of yarn.

The saving analysis has been calculated in terms of fiber cost, dyeing charges, effluent treatment cost after dyeing, energy and water savings due to the absence of yarn dyeing. The effluent generated is taken with the MLR of 1:6 (1 kg of material requires 6 L of water). Effluent is generated in 7 stages of dyeing process namely scouring, hot wash, neutralization, actual dyeing, washing, neutralization and hot wash. Apart from this there are other sustainability benefits that include reduction of green house gases liberated and pollution of water bodies.

From the above table the savings in fabric is evident since virgin cotton is expensive when compared to recycled cotton. Moreover highest savings is evident in the 100% recycled cotton. As the recycled cotton yarn is blue in color there is savings in water, energy, dyes and auxiliaries and effluent treatment cost (Hawley 2017; Harmony 2017; Chavan 2014; LeBlanc 2017).

Table 12 Cost for raw material and yarn development from virgin cotton and virgin cotton/recycled cotton blends

S. No.	Particulars	100% cotton	100% RC	80:20 RC/VC	60:40 RC/VC	40:60 RC/VC
1	Raw material Denim cutting waste/virgin cotton	–	24	19 + 36 = 55	14 + 72 = 86	10 + 108 = 118
2	Fiber extraction cost	–	40	32	23	17
	Total Fiber cost	**180**	**64**	**87**	**109**	**135**
3	Yarn development/kg	10	10	10	10	10
	Total per kg (commercial cost: fiber + yarn)	**190**	**74**	**97**	**119**	**145**
	Total per kg (sample development charge)	**5160**	**5054**	**5077**	**5099**	**5125**
	Costing for 1000 m of fabric					
4	Cost for 125 kg of yarn (the requirement for 1000 m of fabric)	23,750	9250	12,125	14,875	18,125
5	Weaving cost for 1000 m @ Rs. 9/m	9000	9000	9000	9000	9000
6	Desizing cost	10	10	10	10	10
	Total cost of 1000 m of fabric (4 + 5 + 6)	32,760	18,270	21,135	23,885	27,135
	Total Cost for 1 m of fabric [(4 + 5 + 6)/1000]	**32.76**	**18.27**	**21.14**	**23.88**	**27.13**

The costing is subject to change; this estimate is as per the cost in the financial year 2014–2015

5 Conclusion

The results focus that recycled fibers can be used for fabric development suitable for apparels. In all the properties analyzed the yarns and fabrics developed from recycled fibers were slightly lower to the yarns and fabrics developed from virgin cotton fibers. To compensate for the loss a small portion of virgin fibers could be blended to give a yarn and fabric almost equal to virgin cotton yarns and fabrics. From the study the blend of recycled yarn 80:20 recycled cotton: virgin cotton is recommended for industrial production as it has the qualities almost equal to yarns made from virgin cotton.

Further savings is evident in this process in all grounds right from raw material to all the resources used. This study will also help the environment by avoiding landfill

Table 13 Savings in cost for raw material and yarn development from virgin cotton and virgin cotton/recycled cotton blends

Particulars	**100% cotton**	**100% RC**	Gain or loss % over control	**80:20 RC/VC**	Gain or loss % over control	**60:40 RC/VC**	Gain or loss % over control	**40:60 RC/VC**	Gain or loss % over control
Total for 1000 m of fabric	32,760	18,270	−44.23	21,135	−35.48	23,885	−27.09	27,135	−17.17
Skirt cost @ 20/skirt (666 skirts for 1000 m)	13,320	13,320		13,320		13,320		13,320	
Total cost for 666 skirts	46,080	31,590	−31.45	34,455	−25.23	37,205	−19.26	40,455	−12.21
Cost per skirt	69.19	47.43		51.73		55.86		60.74	

Table 14 Savings in cost: virgin cotton fabric versus recycled cotton fabric

S. No.	Particulars	Remarks	Savings	Actual savings for 1000 m in rupees
1	Fibre cost	VC-RC (180-64)	116/kg	3132
2	Dyeing cost	dyeing is not required as the recycled yarn is blue in colour	50/kg	6250
3	Effluent cost MLR[a] = 1:6 i.e. 42 lit/kg	Since dyeing is not necessary effluent treatment charges is saved	20 p/lt	1050
4	Energy cost	Since dyeing is not necessary energy charges is saved	12/kg	1500
5	Water cost	Water is purchased for dyeing in Tirupur, Tamilnadu	Rs. 750/1000 L of water	3937
	Total savings for 1000 m of fabric			**15,869**

[a]Material liquor ratio

dumping and reuse of the waste from apparel production. The Indian garment industry is poised to achieve a target of $25 billion as per the estimate of AEPC and leading retail giants all over the world are looking out for new marketing claims to woo their customers. One such tool to catch the attention of the consumer is the presence of recycled materials in their products with the aim to move towards sustainability. Each industry is competing with their counterparts to showcase their products in a better way in the eye of the consumer and their move towards sustainable production would help in catching the attention of the target market to make the purchase for better promotions and business prospects. The change in production is already in vogue and great care is taken to utilize every ounce of materials and resources from design, prototype production, bulk production and end of life management of products. The trend is not reuse and recycling but the extent to which recycled materials are used and benefits derived out of this symbiosis. To quote,

> 'The purpose – where I start – is the idea of use. It is not recycling. It is reuse'
>
> —Issey Miyake (Brainy Quote 2017)
>
> 'I only feel angry when I see people throwing away things we could use'
>
> —Mother Teresa (Brainy Quote 2017)

Acknowledgement The authors wish to acknowledge the support from Ms. **K.G. Denim**, Coimbatore, **The Management**, **Kumaraguru College of Technology**, Coimbatore and **KCT-TIFAC CORE**, Kumaraguru College of Technology, Coimbatore.

References

Abou-Nassif AG (2014) A comparative study between physical properties of compact and ring yarn fabrics produced from medium and coarser yarn counts. J Text 2014:1–6

Ahmad SS, Mulyadi IMM, Ibrahim N, Othman AR (2016) The application of recycled textile and innovative spatial design strategies for a recycling centre exhibition space. Procedia Soc Behav Sci 234:525–535. https://ac.els-cdn.com/S1877042816315257/1-s2.0-S1877042816315257-main.pdf?_tid=8a2517c8-e97a-11e7-ae1e-00000aab0f01&acdnat=1514210005_34ca328a633e60df835ab7cee91320a4

Al-Salem SM, Lettieri P, Baeyens J (2009) Recycling and recovery routes of plastic solid waste (PSW): a review. Waste Manag 29:2625–2643

Anonymous (2017) Case study of the textile industry in Tirupur. http://www.roionline.org/books/Industrial%20ecology_chapter05_Tirupur.pdf. Accessed 28 Dec 2017

APCO (2017) Australia Post—2016 outstanding achievement in product stewardship. https://www.packagingcovenant.org.au/documents/item/1086. Accessed 28 Dec 2017

Bagwan ASA, Patil A (2016) Optimization of opening roller speed on properties of open end yarn. J Textile Sci Eng 6:231–234

Barclay S, Buckley C (2000) Waste minimisation guide for the textile industry-a step towards cleaner production. http://www.tex.tuiasi.ro/biblioteca/carti/Articole/Waste_Minimisation_Guide_for_the_Textile_Industry_A_Step_Towards.pdf. Accessed 28 Dec 2017

Barella A, Manich AM, Marino PN, Garófalo J (1973) Factorial studies in rotor-spinning part i: cotton yarns. JTI 74:329–339

Bhalodia NR, Shukla VJ (2011) Antibacterial and antifungal activities from leaf extracts of *Cassia fistula* l.: an ethnomedicinal plant. J Adv Pharm Technol Res 2:104–109

Bhatia D, Sharma A, Malhotra U (2014) Recycled fibers: an overview. IJFTR 4:77–82

Blackburn RS (2015) Sustainable apparel production. processing and recycling. Woodhead Publishing Limited, Cambridge, UK

Brainy Quote (2017) Recycling quotes. https://www.brainyquote.com/topics/recycling. Accessed 31 Dec 2017

Chavan RB (2014) Environmental sustainability through textile recycling. J Textile Sci Eng S2:007

Circle Economy (2017) Closing the loop: 3 case studies highlighting the potential impact of high-value, textile recycling. https://www.circle-economy.com/closing-the-loop-3-case-studies-highlighting-the-potential-impact-of-high-value-textile-recycling/. Accessed 28 Dec 2017

Claudio L (2007) Waste couture: environmental impact of the clothing industry. Environ Health Perpect 15:A 449–A 454

Crighton KN (1993) Unbleached denim finds new life in blue jean paper products. Tappi J 76:41–42

Crul M, Diehl JC (2006) Design for Sustainability—a practical approach for developing economies. UNEP. http://www.unep.fr/shared/publications/pdf/dtix0826xpa-d4sapproachen.pdf

EPA (1996) Best management practices for pollution prevention in the textile industry. http://infohouse.p2ric.org/ref/02/01099/0109901.pdf. Accessed 28 Dec 2017

ETBPP (1999) How to profit from less waste and lower energy use in the textiles industry-textiles workbook-ET184. http://infohouse.p2ric.org/ref/23/22045.pdf. Accessed 28 Dec 2017

Faud-Lake A (2009) Design activism: beautiful strangeness for a sustainable world. Earthscan, UK. https://designopendata.files.wordpress.com/2014/05/designactivism-beautifulstrangenessforasustainableworld_alastairfuadluke.pdf

Fiber2fashion (2017) Post-consumer waste recycling in Textiles. http://www.fibre2fashion.com/industry-article/6901/post-consumer-waste-recycling-in-textiles. Accessed 29 Dec 2017

Fletcher K (2008) Sustainable fashion and textiles: design journeys. Earthscan, UK

Frost J (2016) Understanding analysis of variance (Anova) and the F-test (18 May). http://blog.minitab.com/blog/adventures-in-statistics-2/understanding-analysis-of-variance-anova-and-the-f-test. Accessed 31 Dec 2017

Gulich B (2006) Designing textile products that are easy to recycle. Recycling in textiles. Woodhead Publishing, UK

Harmony (2017) Clothing and textile recycling. https://harmony1.com/clothing-textile-recycling/. Accessed 31 Dec 2017
Hawley JM (2006) Digging for diamonds: a conceptual framework for understanding reclaimed textile products. Cloth & Textiles Res J 24:262–275
Hawley JM, Sullivan P, Kyung-Kim Y (2017) http://fashion-history.lovetoknow.com/fabrics-fibers/recycled-textiles. Accessed 31 Dec 2017
Hearle JWS, Lord PR, Senturk N (2008) Fibre migration in open-end-spun yarns. J TEXT I 63:605–607
Hickman M (2010) Pre-or post-consumer recycled content? (5 May). http://edition.cnn.com/2010/LIVING/wayoflife/05/05/pre.post.consumer.recycling/index.html. Accessed 29 Dec 2017
Ichim M, Sava C (2016) Study on recycling cotton fabric scraps into yarns. Buletinul AGIR nr 3:65–68. http://www.agir.ro/buletine/2705.pdf. Accessed 30 Dec 2017
ITJ (2008) Fabric usage and various fabric losses in cutting room. http://www.indiantextilejournal.com/articles/FAdetails.asp?id=1307. Accessed 30 Dec 2017
Jankauskaite V, Macijauskas G, Lygaitis R (2008) Polyethylene terephthalate waste recycling and application possibilities: a review. Mat Sci 14:119–128
Jenkin M (2015) 11 things we learned about achieving a zero-waste fashion industry (15 January). https://www.theguardian.com/sustainable-business/sustainable-fashion-blog/2015/jan/14/10-things-learned-zero-waste-fashion-industry. Accessed 29 Dec 2017
Joshi M (2001) Environment management systems for the textile industry: a case study. IJFTR 26:33–39. https://pdfs.semanticscholar.org/d17f/a01913527acc420590eeebf9f15352c3226d.pdf. Accessed 28 Dec 2017
Kaur B, Balgir PP, Mittu B, Singh H, Kumar B, Garg N (2012) Comparison of antimicrobial susceptibility of bacteriocins from lactic acid bacteria with various antibiotics against *Gardnerella Vaginalis*. Asian J Pharm Clin Res 5:179–181
Denim KG (2017) Products. http://www.kgdenim.com/products/. Accessed 29 Dec 2017
Lane DM (2017) Analysis of variance—introduction (Chap. 15). onlinestatbook.com/2/analysis_of_variance/intro.html. Accessed 31 Dec 2017
Lau Y-I (2015) Reusing pre-consumer textile waste. SpringerPlus (27 Nov). https://www.ncbi.nlm.nih.gov/pmc/articles/PMC4796196/. Accessed 29 Dec 2017
LeBlanc R (2017) What is textile recycling (1 March). https://www.thebalance.com/the-basics-of-recycling-clothing-and-other-textiles-2877780. Accessed 31 Dec 2017
Leonas KK (2017) The use of recycled fibers in fashion and home products. In: Muthu S (eds) Textiles and clothing sustainability. Textile science and clothing technology. Springer, Singapore
@@@Lew D (2017) Industrial ecology (April 27). https://www.drdarrinlew.us/industrial-ecology/concrete-examples-three-case-studies.html. Accessed 28 Dec 2017. Manufacturing: an overview. Fash text. 3:1–16
Moore SB, Ausley LW (2004) Systems thinking and green chemistry in the textile industry: concepts, technologies and benefits. J Clean Prod 12:585–601
Nayak R, Padhye R (2013) The use of laser in garment
Necef, Kurtoğlu O, Necdet S, Maşuk P (2013) A study on recycling the fabric scraps in apparel manufacturing industry. J Text Apparel 23:286–289
Niinimäki K, Hassi L (2011) Emerging design strategies in sustainable production and consumption of textiles and clothing. J Clean Prod 19:1876–1883
Niinimäki K (2006) Ecodesign and Textiles. RJTA 10:67–75
NPTEL (2013) New spinning systems (July). http://nptel.ac.in/courses/116102038/32. Accessed 30 Dec 2017
Online Textile Academy (2017) Ring spinning vs. open end spinning. https://www.onlinetextileacademy.com/2017/11/ring-vs-open-end-spinning.html. Accessed 31 Dec 2017
Papanek V (1995) The green imperative: ecology and ethics in design and architecture. Thames and Hudson Ltd., London, UK
Parellax Limited (2017) How to stop waste in the garment factory. http://www.fibre2fashion.com/industry-article/2971/how-to-stop-waste-in-a-garment-factory. Accessed 29 Dec 2017

Reller LB, Weinstein M, Jorgensen JH, Ferraro MJ (2009) Antimicrobial susceptibility testing: a review of general principles and contemporary practices. Clin Infect Dis 49:1749–1755

Rouse M (2017) Statistical analysis. http://whatis.techtarget.com/definition/statistical-analysis. Accessed 31 Dec 2017

Rowe RG (2000) Textile recycling machine. US Patent No. US6061876 A. https://www.google.com/patents/US6061876. Accessed 29 Dec 2017

Roznev A, Puzakova E, Akpedeye F, Sillstén I, Dele O, Ilori O (2017 Recycling in textiles (19 April). http://www5.hamk.fi/arkisto/portal/page/portal/HAMKJulkisetDokumentit/Tutkimus_ja_kehitys/HAMKin%20hankkeet/velog/TEXTILE_RECYCLING3.pdf. Accessed 23 Dec 2017

Schmidt M, Nakajima M (2013) Material flow cost accounting as an approach to improve resource efficiency in manufacturing companies. Resources 2:358–369

Scraps into yarns. Buletinul AGIR nr 3:65–68. http://www.agir.ro/buletine/2705.pdf. Accessed 30 Dec 2017

Sharma R, Goel A (2017) Development of nonwoven fabric from recycled fibers. J Textile Sci Eng 7:289–292

Sultan MA, El-Hawary IA (1974) A comparison of the properties of open-end-spun and ring-spun yarns produced from two egyptian cottons. J TEXT I 65:194–199

Talyq (2017) Drafting a A-line skirt (dart or no darts?). http://artisanssquare.com/sg/index.php?topic=588.0. Accessed 31 Dec 2017

Tanapongpipat A, Khamman C, Pruksathorm K, Hunsom M (2008) Process modification in the scouring process of textile industry. J Clean Prod 16:152–158

Taşdemır M, Koçak D, Usta I, Akalin M, Merdan N (2007) Properties of polypropylene composite produced with silk and cotton fiber waste as reinforcement. Int J Polym Mater 56:1155–1165

Telli A, Babaarslan O (2017) Usage of recycled cotton and polyester fibers for sustainable staple yarn technology. TEKSTİL ve KONFEKSİYON 27:224–233. http://dergipark.gov.tr/download/article-file/345775. Accessed 31 Dec 2017

Textile School (2017) Open end spinning. http://www.textileschool.com/articles/112/open-end-spinning. Accessed 31 Dec 2017

Toprak T (2017) Textile industry environmental effects and approaching cleaner production and sustainability: an overview. J Textile Eng Fashion Technol 2:1–16

UNEP (1996) Cleaner production—a training resource package. http://www.uneptie.org/shared/publications/pdf/WEBx0029xPA-CPtraining.pdf. Accessed 28 Dec 2017

Vardhman (2012) Vardhman A&E threads installs zero liquid discharge ETP in Perundurai, Tamil Nadu, India (2 November). http://www.amefird.com/in/vardhman-a-e-threads-installs-zero-liquid-discharge-etp-in-perundurai-tamil-nadu-india. Accessed 23 Dec 2017

Wanassi B, Azzouz B, Hassen MB (2015) Recycling of post-industrial cotton wastes: quality and rotor spinning of reclaimed fibers. Int J Adv Res (Indore) 3:94–103

Wang Y (2010) Fiber and textile waste utilization. Waste Biomass Valor 1:135–143

Well made Clothes (2017) Why pre-consumer recycled cotton is really sustainable. https://wellmadeclothes.com/articles/PreConsumerRecycledCottonIsMoreSustainable/. Accessed 29 Dec 2017

Wenzel H, Knudsen HH, Sójka-Ledakowicz J, Machnowski V, Koprowska J, Grzywacz K, Hansen J, Birch H, Pedersen BM, Jozwik A (1999) Cleaner technology transfer to the polish textile industry. https://pdfs.semanticscholar.org/ae78/59d44fd74f60681fcb9add1499e790a3eba1.pdf. Accessed 28 Dec 2017

Yu C (1999) Open-end spinning using air-jet twisting. TRJ 69:535–538

Printed in the United States
By Bookmasters